P9-DIZ-843

STATE OF THE UNIVERSE 2007

Martin Ratcliffe

STATE OF THE UNIVERSE 2007

NEW IMAGES, DISCOVERIES AND EVENTS

Published in association with

Martin Ratcliffe FRAS
Wichita
Kansas
USA

Front cover illustration: In January 2002, a dull star in an obscure constellation suddenly became 600,000 times more luminous than our Sun, temporarily making it the brightest star in our Galaxy. The mysterious star, known as V838 Monocerotis, has long since faded back to obscurity, but this Hubble Space Telescope image reveals dramatic changes in the illumination of surrounding dusty cloud structures. The effect, called a 'light echo', has been unveiling never-before-seen dust patterns since the star suddenly brightened. Image courtesy NASA, ESA, and The Hubble Heritage Team (STScI/AURA).

Back cover illustration: This infrared image of the spiral arms of the nearby galaxy Messier 81 was obtained by NASA's Spitzer Space Telescope. Winding outwards from the bluish-white central bulge of the galaxy, where old stars predominate and there is little dust, the grand spiral arms are dominated by infrared emission from dust. The infrared-bright knots within the spiral arms show where massive stars are being born in giant clouds of ionized gas and dust. Image courtesy NASA/JPL/Caltech/Harvard-Smithsonian Center for Astrophysics.

SPRINGER-PRAXIS BOOKS IN POPULAR ASTRONOMY
SUBJECT ADVISORY EDITOR: John Mason B.Sc., M.Sc., Ph.D.

ISBN 10: 0-387-34178-1 Springer Berlin Heidelberg New York

Springer is a part of Springer Science + Business Media *(springeronline.com)*

Library of Congress Control Number: 2006929599

Cover design and cartoons: Jim Wilkie
Typesetting and design: BookEns Ltd, Royston, Herts., UK

Printed in Germany on acid-free paper

CONTENTS

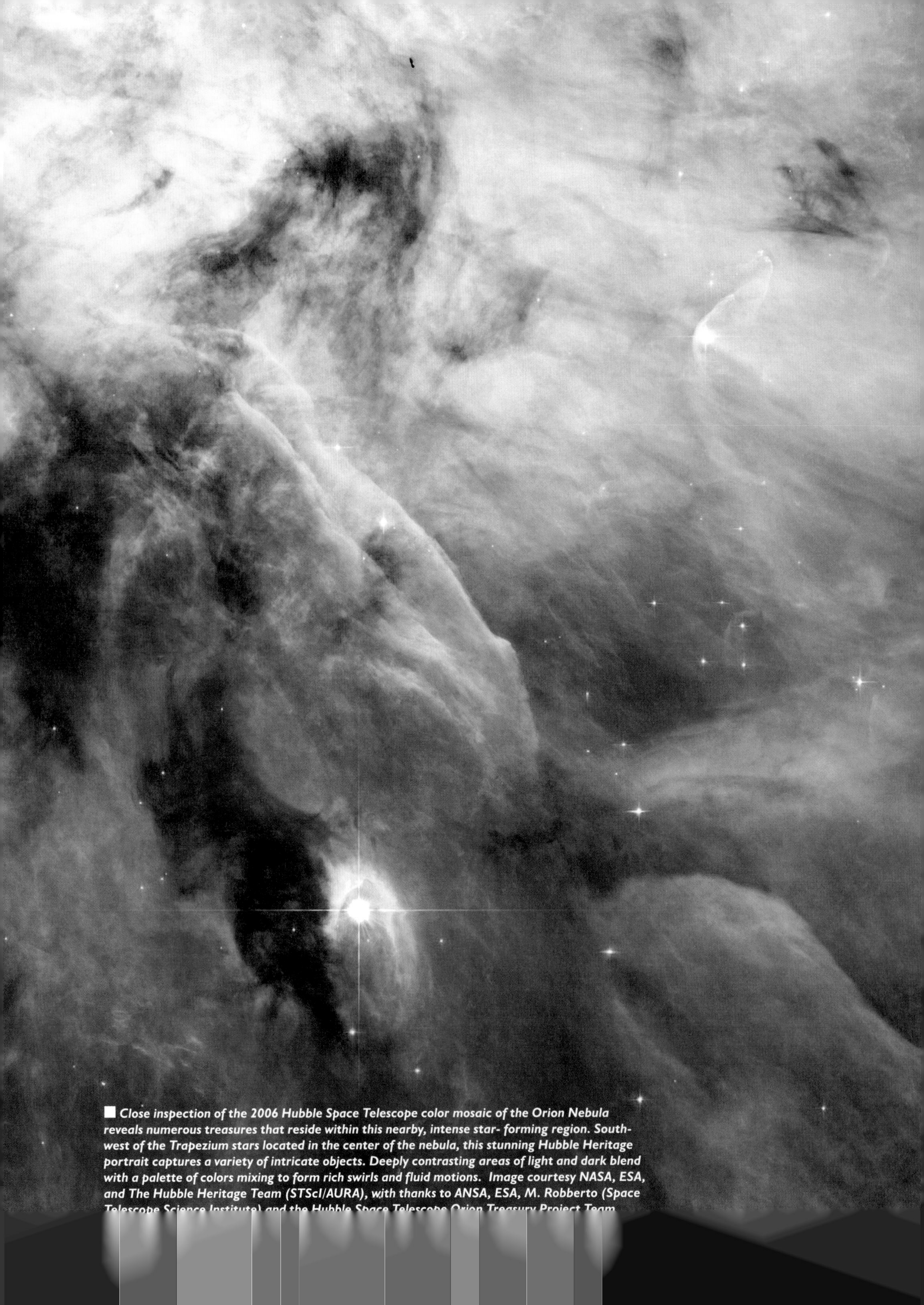

■ *Close inspection of the 2006 Hubble Space Telescope color mosaic of the Orion Nebula reveals numerous treasures that reside within this nearby, intense star- forming region. Southwest of the Trapezium stars located in the center of the nebula, this stunning Hubble Heritage portrait captures a variety of intricate objects. Deeply contrasting areas of light and dark blend with a palette of colors mixing to form rich swirls and fluid motions. Image courtesy NASA, ESA, and The Hubble Heritage Team (STScI/AURA), with thanks to ANSA, ESA, M. Robberto (Space Telescope Science Institute) and the Hubble Space Telescope Orion Treasury Project Team.*

Preface

STATE OF THE UNIVERSE 2007

IMAGINE, IF you will, standing next to a spectacular river flowing down a mountainside. High in the distance, at a higher altitude, there's a cascade of water heading in your direction. Right where you are standing, and taking your photograph, the river is broad, turbulent, and full of drama. Near the river bank, some water even turns back upstream in small eddies. These momentary diversions are quickly whisked downstream by the full flow of the river. There are many small ripples, some large waves, and here and there, the infrequent but spectacular explosion of spray.

You take your photograph. It's a snapshot, a still scene, the frenzied motion of the river frozen for a moment in time. The river is broad from your vantage point, yet apparently stationary in your image.

This book is like your photograph. It's a snapshot, one moment in time, of a veritable gush of astronomical information that is pouring down from high mountain top observatories, and space telescopes orbiting high above the Earth.

Within its pages is an attempt to capture some of the flavor of the dynamic, fast-paced, and sometimes turbulent, flood of astronomical information about our Universe. It is an attempt to provide an impression, like your photograph, of our state of knowledge of the Universe. Like each passing wave in the river, the information in this book is not static, it's moving, and by the time you read this, in places, it will have already moved on.

This is the first book in an annual series that brings you the some of the greatest discoveries in astronomy that have occurred during the previous year. It's a bridge between the public pronouncements of astronomy news and the professional researcher. With 30-second sound bites about a recent Hubble discovery, many listeners may be left wanting, knowing there is more to the story than the popular media have time for.

New extrasolar planets, new views of our Milky Way, evidence of black holes colliding, and the most dramatic images from the world's best telescopes, are the most visible part of a deeper story. It's an astounding story that's being revealed, step by step, through careful, painstaking work by the world's astronomers. By reading this book, you will gain a deeper understanding of what the images mean and how they tie in to the broader picture of astronomical research.

Telescopes in orbit, such as Spitzer, Hubble, Swift, Chandra and XMM Newton, are reaching far beyond any telescope before them. Ground-based optical and radio telescopes are providing new and unprecedented views of the depths of space. The Very Large Telescope (VLT), Keck, Gemini, the Very Large Array (VLA) and the Green Bank Radio Telescope are just a few of the instruments that are revolutionizing our view of the cosmos.

Large scale deep surveys, notably the Sloan Digital Sky Survey, are producing astonishing new results, from the recognition of new streams of stars in our own Milky Way, out to some of the most distant quasars known. New telescopes soon to come on line, or in the early stages of planning, such as the new Large Binocular Telescope, or the futuristic Giant Magellan Telescope, indicate that the flood of information is set to increase dramatically. Strange-looking telescopes search for gravitational waves (LIGO), and the origin of cosmic rays (VERITAS). Each new telescope is historically associated with giant leaps forward in our understanding of the cosmos.

Keeping track of all the new results is hard. The exciting developments in the understanding of our origins, of the early beginnings of the Universe, clues to the nature of dark matter and dark energy, of how planets are formed, and how stars live out their lives and die, occur every month. Each new result adds a tiny piece to the jigsaw puzzle, leading the way to a fuller and more complete understanding of the Universe around us. Rarely are such details offered in one place. This book attempts to fill the gap between research and everyday news. It's a unique insight into many of these developments, not only from the news releases from major observatories, but also from the researchers themselves.

The book is split into two main parts. The first part, 'A Year in News and Pictures', reviews month-by-month some of the major new findings in chronological order covering the period from April 2005 to March 2006 (defined by the publication schedule). While there is no way one can review every news story (such a book would run to over 500 pages), the aim is to provide a broad survey of the leading stories that reflect not only the picturesque and dramatic, but also the astronomically significant and astonishing.

Each news item has a number of web links to enable you to take your research further. The links take you to web sites designed for the general public reader, and I have provided, in many cases, additional links to the actual research paper or research web site. These hard-to-find web sites provide much deeper information for those who have a background in physics and astronomy and wish to read further. The research papers are often listed on a web site called 'astro-ph', short for astrophysics. This is both a goldmine for undergraduate students studying astronomy at major universities, while at the same time an opaque reservoir of inexplicable titles for those not well versed in physics and astronomy research. If that is the case, stick to the first web site listed as a link.

During the research for this book I've consulted the extensive volume of press releases available from all the major observatories around the world, and with that significant thanks go to all the press officers and education and public outreach specialists of all astronomical institutions. Their work is vital in conveying the complex language of research into the exciting and palatable words for public consumption.

The second part of this book consists of invited review articles from leading researchers and science writers. I am enormously indebted to them for their major contributions to this volume. The topics highlight some of the most active areas of current research, reflecting the state of our knowledge now, and placing many of the news stories in the first section of the book into a broader context.

I am very grateful to James Kaler for writing a review of the year 2005-06 for this book. He is well known in the planetarium community for keeping many of its profession up to date each year through lectures at annual conferences, and he was a natural choice for writing about our current 'State of the Universe'. His article highlights the most significant discoveries of the year, and places them in context of our broader astronomical knowledge.

The remaining nine chapters in the features section focus on specific topics currently undergoing the most active research. Ray Villard, a science writer in his own right and Hubble's Press Officer since the launch of the Hubble Space Telescope, brings us the highlights of the year from that most magnificent of instruments orbiting a few hundred kilometers above our heads.

Richard McCray, world-renowned expert on the famous 1987 supernova in the Large Magellanic Cloud, gives us the most recent update regarding the search for the central compact object, and the dramatic flaring of the gaseous ring surrounding the supernova. The collision of the supernova debris with this slow-moving ring is of gargantuan proportions, and McCray's review is timely with the entire ring now illuminated.

Neil Gehrels, Principal Investigator for the SWIFT gamma-ray observatory, with co-author and colleague, Peter Leonard, outline some dramatic new discoveries and the progress made in the past year in our understanding of the origin of some of these bursts.

The Milky Way was a hot topic at the January 2006 Winter AAS meeting, with many new results from the Spitzer Infrared Space Telescope, and from the SDSS. Chris Wanjek, an experienced science writer, provides a dramatic overview of the most recent results, from the central bar to a warped disk.

The director of the newly completed Large Binocular Telescope (LBT), Richard Green, and colleague John Hill, provide insightful commentary on this revolutionary telescope.

■ *Editor Martin Ratcliffe at Wichita State University's Lake Afton Public Observatory, writes about astronomy and teaches an introductory course in Cosmology at the WSU. He was Director of the first digital planetarium theater in Wichita, Kansas. The Lake Afton facility is like many around the world that offer opportunities for the general public and students to find out more about astronomy. Image courtesy Classic Impressions by Trey Allen, with acknowledgement to Lake Afton Public Observatory.*

The pair of gravitational wave observatories called LIGO recently began a continuous run of observations. Laura Cadonati gives us the benefit of her considerable expertise to tell the story of this remarkable development. The potential for great discoveries probably lies in the next few years. Read her article to see what we might expect.

Michelle Thaller, an expert at conveying information to the public in an exciting way, as well as being an astronomer, reviews the outstanding science coming from the infrared Spitzer Space Telescope. The images are dramatic. The science is astounding. Check out what's cool with Spitzer this year.

One of the great surveys of the cosmos using the Hubble Space Telescope is the COSMOS galaxy survey. Anton Koekemoer, an astronomer at the Space Telescope Science Institute, leads the observations, and we are privileged to have his contribution. News from COSMOS graces the halls of every recent AAS meeting. Anton provides a complete background to the survey.

Finally, on a lighter, yet serious, note, Phil Plait provides humorous and incisive critique of the crazy claims that appear from time to time. Past stories like the wild face on Mars, or the Moon Hoax, rear their heads, and typically require some simple logic to discredit. Phil dedicates his web site badastronomy.com, to exposing these wild claims for what they are.

Finally, you'll find appendices filled with information about all the large telescopes, current and planned, with web links to keep you abreast of their current status and most recent breaking news, in addition to a list of extrasolar planets. These lists will be updated each year.

This book is designed to begin a new level of exploration into astronomy. The general reader interested in astronomy, students at community colleges and first year undergraduates, teachers and writers will all find valuable information here.

If you are a student of astronomy at a local community college and need some ideas and access to information for term papers, this book provides a very concise doorway to the most recent and exciting discoveries in astronomy, and leads you directly to the web sites that provide images and technical information you can use for your paper. For more advanced students, direct links to the actual research papers written by the astronomers themselves open the door to second and third year astronomy and physics students with a strong mathematical background.

Amateur astronomy clubs will find this useful as a source for potential speakers via the topical news items.

More advanced students who may be figuring out what area of research interests them as they pursue a career will be able to gauge which universities are studying particular areas of astronomy from their news releases and research team web sites.

For the general reader wanting to keep pace with developments in the science of astronomy, this book is probably all you need, but passionate followers of astronomy will also subscribe to some of the fine monthly magazines that regularly review astronomy, such as *Sky and Telescope* and *Astronomy* magazines.

For the science writer, this book provides a handy reference for all those press releases you wonder what to do with. If you want to know what stories broke in June 2005, this book provides easy access to the main stories, and the web links to help you research further. This is especially valuable to writers who want to link various 'discoveries' into the context of a bigger picture story.

To help guide you along in further discovery, if you come across a word you don't understand, like "magnetar", try entering the word into a search on the internet to find its meaning. Look for links to bona-fide university web sites to delve deeper into the topic of interest.

Thanks certainly go to the team of science writers whose work provided the raw material for the news items. The people listed below were particularly useful, if not directly, then indirectly.

My thanks go to Lori Stiles of the University of Arizona News Services, Megan Watzke (Chandra PIO), Peter Michaud (Gemini PIO), Lars Lindberg Christensen (European Hubble Space Telescope PIO), Laura Kraft (Keck Observatory PIO), David Aguilar (Harvard University), Christine Pulliam (Harvard University), Whitney Clavin (JPL),

Dolores Beasley (NASA headquarters), Cheryl S. Gundy and Donna Weaver, (Space Telescope science writers), Susan Hendrix (Goddard Space Flight Center), Steve Koppes University of Chicago, Robert Tindol (Caltech), Tina McDowell (Carnegie Institution), and last but not least, Rebecca Johnson (McDonald Observatory PIO).

It is important to remember one thing regarding press releases. They do provide a popular level introduction to cutting edge research. However, there have been instances where a new discovery or claim has been proved incorrect with further research. Such is the natural process of science. This is the strength of science at work.

Typically, new research takes years to ferret out the true story. A press release highlights one particular nugget of information on the way to this fuller understanding. Therefore, each news item in this book should be read with this in mind. If you accept any news story as the correct view, you may find an opposing view held by other scientists. The debate between them is competitive and, at times, fierce. Experiencing this process is to become engaged in it.

Many of the dramatic steps forward in our understanding of the Universe never make the pages of local newspapers, and the general public and students miss a great opportunity to share in the excitement of this rapidly growing subject. After all, astronomy encompasses the entire Universe. Our lives are intimately linked to the formation of stars, since we are made of the chemical elements that are known to be formed inside massive stars. Earlier generations of massive stars have long since exploded, seeding the universe with elements heavier than hydrogen and helium. Iron, an element that courses through our veins in the hemoglobin molecule in blood, originally came from the centers of early stars. When you look at the sky, there's little surprise that most of us feel some sense of awe – it's more than awe, it's a direct connection we have to the Universe. It's in our blood. As I tell my students, if you don't feel that connection, you're probably not alive.

And my thanks would not be complete without mentioning my students who have passed through the astronomy introductory courses at Butler County Community College, Baker University, and the cosmology course at Wichita State University, all in Kansas. Passing on the free access to the world of astronomy is a passion I enjoy sharing, and it is my hope that many students will carry the interest for life.

Unknown to my editor and publisher, this kind of book has been in the back of my mind ever since the first day I attended a meeting of the American Astronomical Society. That was in 1992 in Columbus, Ohio. As a planetarium director for the Buhl Planetarium at Pittsburgh's Carnegie Science Center, I was keen to ensure our shows reflected current research. On arrival at the meeting I met Steve Maran, Press Officer of the AAS, and he graciously allowed me access to the press room. Ever since then, missing only a few meetings, I have witnessed the veritable avalanche of exciting astronomical results being discussed and published. Each year I really expected that the next year could not bring any more startling results. And yet each year always does. My sincere thanks and appreciation goes to Steve Maran for opening that door to me 14 years ago.

Soon after my first AAS meeting, I began to think about instituting an annual 'State of the Universe' address in January around the time of the January AAS meeting. While this never came to fruition, I have kept every news release from those days in the hope that someday, items like that would be of some use. This book is the brainchild of John Mason and Clive Horwood, my patient and ever-supporting Editor and Publisher. Yet when they suggested the book to me, their ideas were falling on already fertile ground. I'd like to thank them profusely for their attentive nurturing, positive encouragement, and patience, for this first volume.

Finally my heartfelt thanks and dedication of my work go to my wife, Shawn, who shares my passion for love and life, and whose active encouragement of my writing enabled this project to be completed, and to the memory of Rev. Dr Gary Cox, who shared in the excitement of astronomical discoveries over the past few years.

Martin Ratcliffe

Wichita, Kansas

August 2006

BUNNY
FELINE
ASTRONOMER
KITCHEN
GULP!
MILK

PUBLISHER'S NOTE

IN MY youth I was an avid collector of 'annuals' which, in the UK, consisted of a large-format book containing strip cartoons, short stories and other items and puzzles designed to stimulate the imagination and therefore the learning process.

In developing the range of books in the Praxis imprint *Popular Astronomy* it occurred to me there was a real need for a book for enthusiasts of all ages on popular astronomy. I found a kindred spirit for this idea in Martin Ratcliffe who is also a lover of annuals.

This provided the ideal opportunity to fulfil a longstanding ambition of mine to produce a book - an annual - containing various reviews of new images, discoveries and events in the Universe. And let's illustrate the chapters with a series of one-page cartoons on the various topics, to lighten the learning curve and make the reader smile.

The cartoons show how the very latest telescopes, exciting discoveries and the astronomers behind those discoveries might be seen through the eyes of a cat! This allowed me to have some 'publishing fun' with our cover designer and illustrator, Jim Wilkie, in developing this idea and thus immortalising my wife Jo's incredible 20-year old Russian Blue cat, Bunny, as AstroKat!

Sincere thanks go to Jim's wife, Rachael, for helping to shape and focus the cartoon ideas with Jim and to Arthur Foulser of BookEns, who surpassed the design challenge I set him for the layout of the text. To our intrepid *Popular Astronomy* Advisory Editor, John Mason, a big thank you for his work on the final selection of images and other essential detail.

I've always believed learning should be fun, so for the first time since I started the Praxis imprint over a decade ago, I decided to write a publisher's note to explain the thinking behind this book. To all readers, please enjoy this annual and other volumes in future years. This is the start of a classic series and the very first volume usually becomes a prime collector's item, so you should keep your copy as a treasure after reading it.

Finally, a big thank you to Harry Blom of Springer New York for his enthusiastic support for this project.

Clive J Horwood

This false-color infrared image from NASA's Spitzer Space Telescope reveals bright regions of star-forming activity within the Trifid Nebula. Spitzer uncovered 30 massive embryonic stars and 120 smaller newborn stars throughout the nebula, in both its dark dust lanes and luminous clouds. These stars are visible mainly as yellow or red spots. Embryonic stars are developing stars about to burst into existence. Image courtesy NASA/JPL-Caltech/J. Rho (SSC/Caltech).

1

A year in NEWS...

Each month a flood of new results pours in from the world's observatories. Here *Martin Ratcliffe* reviews the major highlights, selected from the hundreds of news reports released between April 2005 to March 2006. You'll find a concise review of the selected discoveries and web links to images and research articles. Each news item reflects another step toward a more complete and fuller understanding of our Universe.

...and PICTURES

APRIL 2005 - MARCH 2006

APRIL 2005

2 April 2005

First Stars seen in Distant Galaxies

THREE LEADING instruments, two orbiting and one on the ground, have detected light coming from the most distant galaxies ever observed, indicating it comes from the first stars to form when the universe was at a much earlier age.

The Spitzer Space Telescope and the Hubble Space Telescope teamed up with the 10-meter Keck telescope on Hawaii. The Hubble Ultra Deep Field image was used to select the most distant galaxies, and Spitzer and Keck were used to determine their nature.

"We used the images from the Hubble Ultra Deep Field to identify objects likely to be galaxies 95 per cent of the way across the observable Universe," said Andrew Bunker of the University of Exeter, England. "These images are our most sensitive picture of the Universe so far, and they enabled us to discover the faintest objects yet."

The Keck Deep Extragalactic Imager and Multi-Object Spectrometer (DEIMOS) was used to take spectra of the galaxies and they showed a redshift of 6, indicating the light left the galaxy at a time when the universe was only about 700 million years old.

"We proved these galaxies are indeed among the most distant known by using the Keck telescopes to take a spectrum," said Elizabeth Stanway (University of Wisconsin, Madison).

Seeing star formation in the early universe, shortly after the Big Bang, was not expected, but is supported by other observations. Following the Big Bang, the universe was filled with hot expanding gas, and no stars. The end of this so-called Dark Age occurred when the first stars turned on. The picture now forming is that the Dark Ages probably ended between 200 and 500 million years after the Big Bang.

http://www.keckobservatory.org/news/science/spitzer/050402.html

http://arxiv.org/abs/astro-ph/0508271

5 April 2005

First Distance Measurement of a Gamma-Ray Burst Detected by Swift

The first distance measurements to Gamma-Ray Bursts (GRBs) detected by the new orbiting Swift spacecraft have been determined. Swift detected two bursts on March 18 and 19, called GRB 050318 and GRB 050319. Shortly after the announcement of the bursts went out, ground based telescopes swung into action to detect any afterglow. The redshift of the afterglow can lead to a distance determination. For GRB 050318 the distance was about 9.2 billion light years, and for GRB 050319 it was 11.6 billion light years.

"Swift will detect more gamma-ray bursts than any satellite that has come before it, and now will be able to pin down distances to many of these bursts too," said Peter Roming, UVOT Lead Scientist at Penn State. "These two aren't distance record-breakers, but they're certainly from far out there."

Swift was launched in November 2004, and provides very rapid acquisition of the location of a GRB to within a few arcminutes, alerting telescopes on the ground to perform follow-up observations. The goal is to understand what causes GRBs, a long standing puzzle since they were first detected over 30 years ago.

http://swift.gsfc.nasa.gov/docs/swift/news/2005/uvot_redshift_psu.html

http://xxx.lanl.gov/abs/astro-ph/0509060

http://xxx.lanl.gov/abs/astro-ph/0511132

6 April 2005

Era of Galaxy and Black Hole Growth Spurt Discovered

Galactic mergers triggering massive bursts of star formation in the distant universe appear to be rich nurturing grounds for the largest black holes. One of the longest duration exposures with the Chandra X-ray Observatory combined with the submillimeter and optical observations reveal very young galaxies just 3 to 4 billion years after the Big Bang experienced tremendous growth of their central black holes. The results provide the first direct evidence that the mass of stars in a galaxy is linked to the size of its central black hole.

"The extreme distances of these galaxies allow us to look back in time, and take a snapshot of how today's largest galaxies looked when they were producing most of their stars and growing black holes," said David Alexander of the University of Cambridge, UK.

Star formation rates indicated by the submillimeter observations from the James Clerk Maxwell Telescope (JCMT) and optical observations from the 10-m Keck telescope, both located on Mauna Kea in Hawaii, are about 100 times the rate in the Milky Way, or about one star per day. Hubble images show the galaxies observed by JCMT were young merging pairs.

"The Keck observations allowed us to determine that these galaxies were forming their stars at a colossal rate," said Dr. Alexander. "Our detection of X-ray emission with Chandra indicates that their black holes were also growing at the same time. These findings provide direct observational support for the simultaneous growth of large galaxies and their black holes."

Supercomputer simulations by Tiziana Di Matteo of Carnegie Mellon University in Pittsburgh, Pennsylvania, show that the merging of two galaxies drives gas into the central regions of galaxies. The sudden inflow of material raises temperatures and pressures, triggers star formation, and provides ample fuel for feeding a central black hole.

"It is exciting that these recent observations are in good agreement with our simulation. We seem to be converging on a consistent picture of galaxy formation with both observations and theory." said Di Matteo.

http://chandra.harvard.edu/press/05_releases/press_040605.html

http://www.sr.bham.ac.uk/nam2005/pr9.html

http://xxx.lanl.gov/abs/astro-ph/0503453

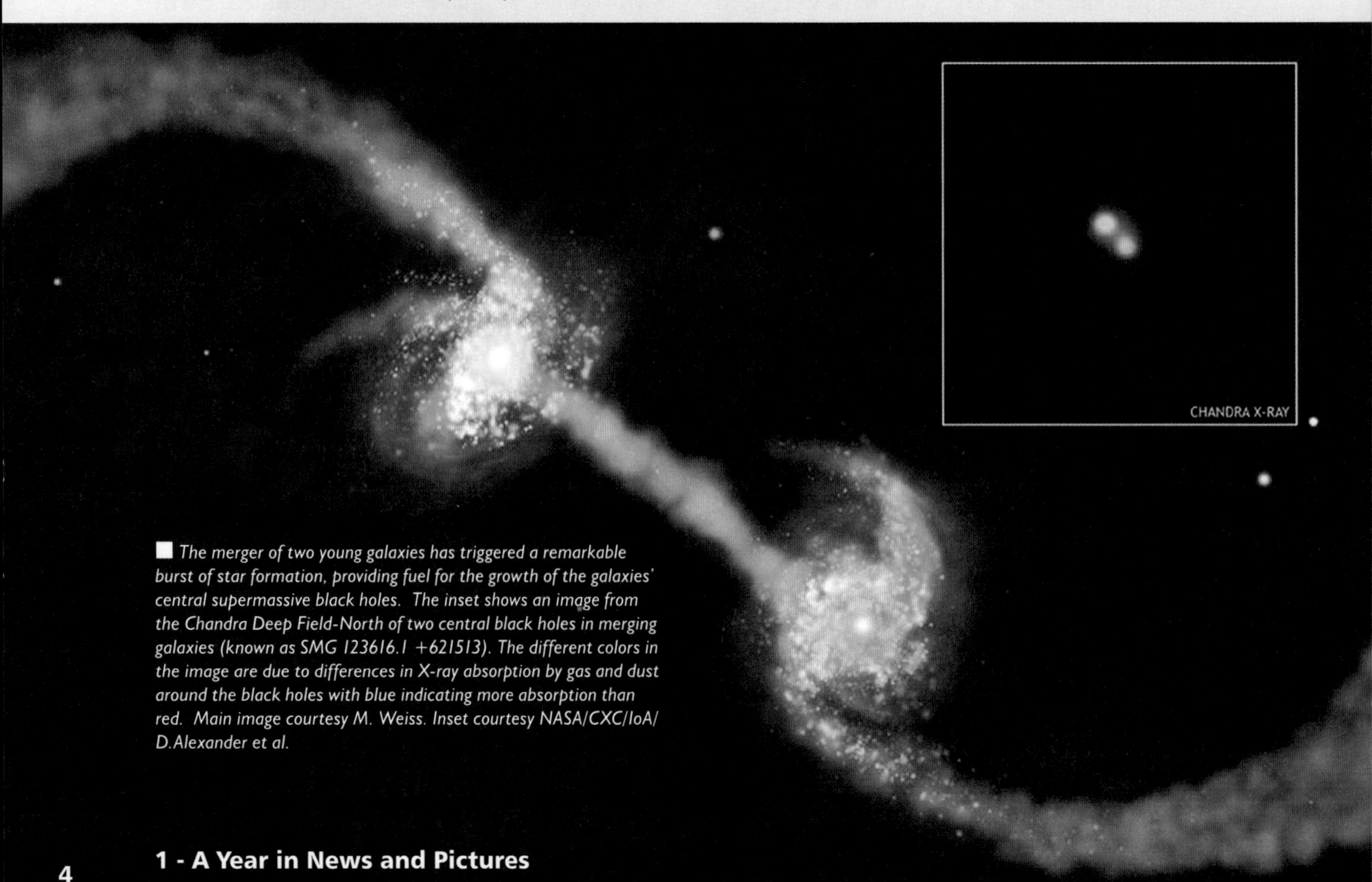

The merger of two young galaxies has triggered a remarkable burst of star formation, providing fuel for the growth of the galaxies' central supermassive black holes. The inset shows an image from the Chandra Deep Field-North of two central black holes in merging galaxies (known as SMG 123616.1 +621513). The different colors in the image are due to differences in X-ray absorption by gas and dust around the black holes with blue indicating more absorption than red. Main image courtesy M. Weiss. Inset courtesy NASA/CXC/IoA/D.Alexander et al.

7 April 2005

Concentrated Dark Matter at the Cores of Fossil Galaxies

Ancient galaxy clusters typically harbor one very massive central galaxy, giving the impression that many of the original cluster members have merged into the single giant galaxy.

Recent X-ray observations of the hot gas surrounding such "fossil clusters" have shown a remarkable concentration of dark and normal matter in the cores of the clusters compared with more normal clusters.

"When we first discovered the large halos of hot gas in which some very compact groups of galaxies are embedded, we realized that just a few billion years of further evolution would leave a single, giant, merged galaxy sitting at the centre of a bright X-ray halo," said Trevor Ponman of the University of Birmingham, England.

Six fossil groups have been identified, most located up to two billion light years away. One of the largest is the giant elliptical galaxy, NGC 6482, located 100 million light years away in the constellation Hercules. This galaxy shines with the equivalent of 110 billion suns. The Chandra and XMM Newton X-ray observatories detected shocked gas in the cluster reaching 10 million degrees C. The heating is explained as coming from the gravitational collapse of the cluster.

Dark matter in the cluster is only detected by its gravitational influence, since it cannot be seen directly. The high temperature leads to a high density of material centered on NGC 6482. To build such structures takes a very long time.

"The explanation for such a centralized dark matter distribution could be that the system formed at very high redshift - when the Universe was very young and dense," said Habib Khosroshahi, also from the Birmingham team.

http://chandra.harvard.edu/press/05_releases/press_040705.html

http://xxx.lanl.gov/abs/astro-ph/0401023

8 April 2005

X-ray Vision of Violence in Interacting Galaxy Clusters

New results from the European Space Agency's orbiting XMM-Newton observatory are illuminating the massive cosmic "pile-ups" that occur between massive galaxies embedded in massive clusters. The gas lying between galaxies is riddled with shock waves that raise its temperature to millions of degrees.

Astronomers used XMM Newton to map the gas distribution and temperature in three giant galaxy clusters and found collisions in clusters occur at up to about 2,000 kilometers per second.

Abell 1750 (A1750) contains two clusters 3 million light years apart just beginning to interact. They are located 1.1 billion light years from Earth. Slightly closer, at 800 million light years from Earth, is Abell 3266. A boomerang-shaped shock wave is caused by a smaller cluster starting to make headway into a more massive cluster. It's expected that the younger A1750 will look like this in a couple of billion years. A third and even older example is A3921, located 1.2 billion light years from Earth. The encounter of two clusters has already occurred, with the lighter cluster almost totally disrupted but leaving a region of shocked gas in its wake.

"This research shows the violent manner by which the largest structures in the Universe form, and that the formation has happened in the recent past," said Elena Belsole of the University of Bristol, England. "The process is still taking place today. In several billion years the group, of which our galaxy is a member, will be torn apart as it merges with the nearby Virgo cluster."

http://xmm.vilspa.esa.es/

http://xxx.lanl.gov/abs/astro-ph/0501377

Above: *Chandra X-ray observations of the giant elliptical galaxy NGC 6482, at the centre of this image, show that it is surrounded by a vast cloud of hot gas (shown in blue), which has a temperature of about 10 million degrees Celsius. This giant galaxy is believed to have grown to its present size by cannibalizing its neighbors, leaving only the X-ray halo to tell the tale. Image courtesy Habib Khosroshahi and University of Birmingham, UK.*

11 April 2005

Scientists Track Collision of Powerful Stellar Winds

The powerful stellar winds of two stars in a binary system are colliding, generating radio waves. The Very Long Baseline Array (VLBA) radio telescope has provided unique details of these winds, providing new insight into the nature of the wind and of the stars themselves.

"The spectacular feature of this system is the region where the stars' winds collide, producing bright radio emission. We have been able to track this collision region as it moves with the orbits of the stars," said Sean Dougherty of the Herzberg Institute for Astrophysics in Canada.

One of the stars is a Wolf-Rayet type, weighing in at 20 times the mass of the Sun. The companion star is more than double this mass, at 50 solar masses. Wolf-Rayet stars are known for their violent stellar winds, and it impacts the lesser wind from the more massive companion. The binary system, called WR 140, has an orbital period of 7.9 years and the elliptical nature of the orbit changes the impact location of the winds and their relative effects on each other. The VLBA observations provide accurate observational data for more detailed modeling of the system.

"People have worked out theoretical models for these collision regions, but the models don't seem to fit what our observations have shown," said Mark Claussen, of the National Radio Astronomy Observatory in Socorro, New Mexico. "The new data on this system should provide the theorists with much better information for refining their models of how Wolf-Rayet stars evolve and how wind-collision regions work."

http://www.nrao.edu/pr/2005/wr140/

http://xxx.lanl.gov/abs/astro-ph/0501391

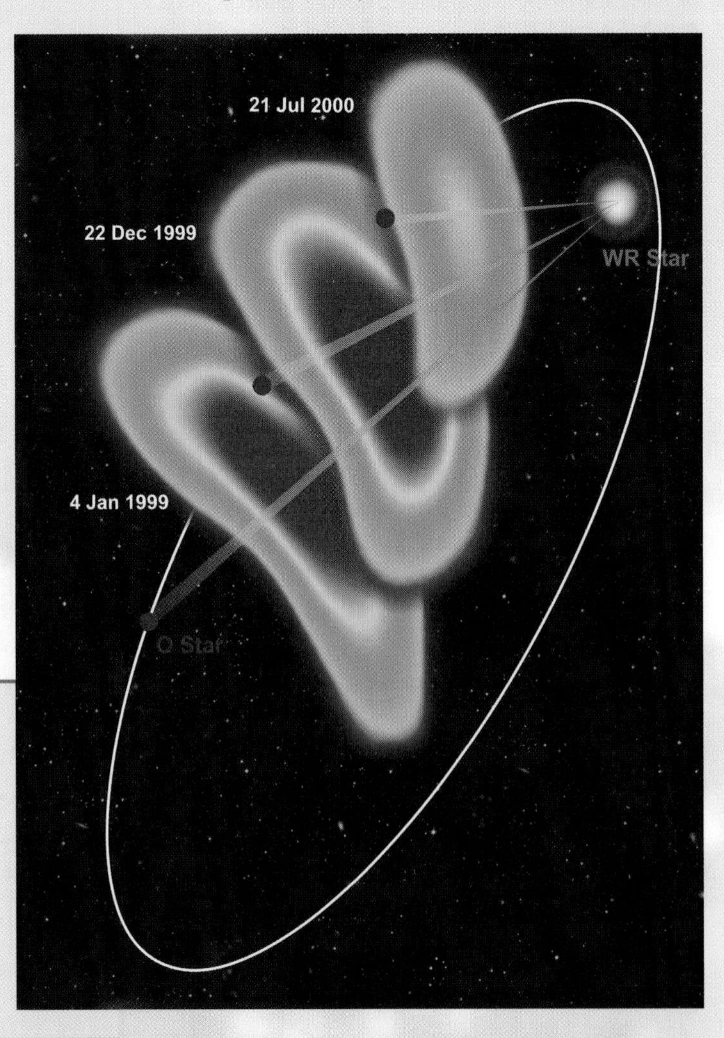

■ ***Left:*** *Motion of the violent collision region where the powerful winds of two giant stars slam into each other. The region moves as the stars, part of a binary pair, orbit one another. This graphic superimposes VLBA images of wind collision region on diagram of orbit of Wolf-Rayet (WR) star and its giant (O-type) companion. Image courtesy Dougherty et al., NRAO/AUI/NSF.*

19 April 2005

Star Explosions on the Half Shell

The question of why some supernovae explosions appear to create a shell from the ashes around them, and yet others do not, has been one of those little troubling problems in astrophysics. The solution, it turns out, is to look long and hard at the supernovae that appear not to have shells.

Recently researchers at the University of Manitoba used the Chandra X-ray Observatory and spent 150 hours imaging one supernova remnant, and to their surprise, a very faint shell was visible that had previously been unseen.

The detection of the shell solves a decades-old mystery. Dr. Samar Safi-Harb and her graduate student, Heather Matheson, say that all star explosions likely create a shell; but some create a faint ('soft') shell, or only half of an easily detectable shell.

"Most star explosions make well-defined and colorful shells, the signature of a classic star explosion," says Safi-Harb. "But even some famous explosions, such as the Crab Nebula, have no obvious shells. It could be that the Crab Nebula is like a soft-shell crab with a thin, barely visible shell."

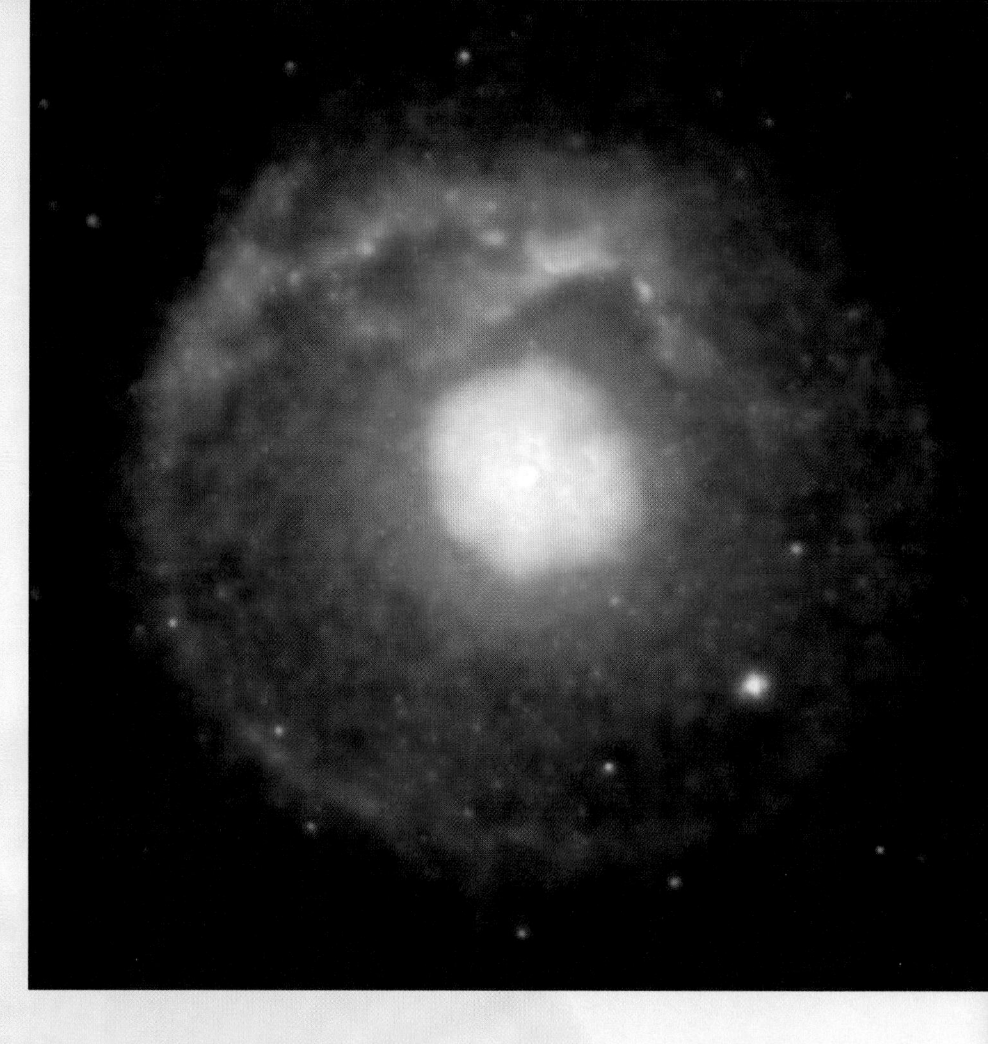

The shells are formed when the heavier elements such as nitrogen, oxygen and iron, produced during the supernova explosion, strike the gas in the interstellar medium. They can exist for thousands of years after the initial explosion. X-rays are produced in these high energy collisions, and understanding such interactions is crucial to understanding how all the heavy elements become distributed throughout the universe. Supernovae are the delivery mechanism for all of the heavy elements that seed the interstellar medium with the raw materials that ultimately will go on to form dusty disks around a new generation of stars. These dusty disks can accrete to form planets. The carbon, nitrogen and oxygen vital for the development of life are, as observations show, scattered throughout interstellar gas clouds thanks to earlier supernovae explosions.

The remnant studied by Safi-Harb and Matheson, called G21.5-0.9, revealed only parts of a shell distributed around the supernova remnant. It's a relatively young remnant at a few thousand years old.

"This is sort of a supernova remnant served on the half-shell," says Matheson. "Why we don't see a full shell as we do around other supernova remnants is the next question we'd like to answer. Why there isn't one around the famous Crab Nebula is another mystery."

Variations in the density of the interstellar medium, and asymmetry in supernova explosions, are two of the likely explanations for the partial shells.

http://chandra.harvard.edu/press/05_releases/press_041905.html

http://xxx.lanl.gov/abs/astro-ph/0504369

25 April 2005

Hubble Celebrates 15th Anniversary with Spectacular New Images

The Hubble Space Telescope, a joint project between the European Space Agency and NASA, reached its 15-year anniversary in orbit, following its launch from the cargo bay of the Space Shuttle on 24 April 1990. To celebrate the event, and to highlight the more than three-quarters of a million images, in addition to the hundreds of thousands spectra, two very large images were released to the public. The two targeted objects are very well known: Eagle Nebula (M16) in Serpens, and Whirlpool Galaxy (M51) in Canes Venatici.

The Eagle Nebula image shows a massive column of dense gas and dust whose outer envelope is glowing by irradiation by ultraviolet light from a group of massive nearby stars.

Above: *This image, made by combining 150 hours of archived Chandra data, shows the supernova remnant G21.5-0.9. The central bright cloud of high-energy electrons is surrounded by a distinctive shell of extremely hot gas. Although many supernovae leave behind bright shells, others do not. This supernova remnant was considered to be one that had no shell until it was revealed by Chandra. Image courtesy NASA/CXC/ Univ.Manitoba/H.Matheson and S.Safi-Harb.*

This image acquired by the Advanced Camera for Surveys on NASA's Hubble Space Telescope shows a soaring pillar of cold gas and dust within a stellar nursery called the Eagle Nebula (M16). The pillar is a giant incubator for newborn stars. A torrent of ultraviolet radiation from a group of massive, hot, young stars [off the top of the image] is steadily eroding the pillar and sculpting fantasy-like landscapes in the gas. Image courtesy NASA, ESA and The Hubble Heritage Team (STScI/AURA).

Such pillars are the nurseries for thousands of new stars. Light from the embedded stars can be detected by infrared radiation, but in this Hubble image they are shrouded by their dusty cocoons hidden in the giant column.

The Whirlpool Galaxy image reveals hundreds of pink glows from HII star-forming regions that delineate the spiral arms, enhanced by spidery dust lanes and massive clusters of hot stars. The companion galaxy that passed through the plane of M51 and now lies behind it has dragged, by gravitational interaction, one of the spiral arms out of place.

The new images celebrate the major list of scientific achievements of the Hubble Space Telescope:

- **Helped astronomers calculate the precise age of the universe (13.7 billion years old);**
- **Helped confirm the existence of a strange form of energy called dark energy;**
- **Detected small proto-galaxies that emitted their light when the universe was less than a billion years old;**
- **Proved the existence of super-massive black holes;**
- **Provided sharp views of a comet hitting Jupiter;**
- **Showed that the process of forming planetary systems is common throughout the galaxy;**
- **Taken more than 700,000 snapshots of celestial objects such as galaxies, dying stars, and giant gas clouds where stars are born.**

http://hubblesite.org/news/2005/12
http://heritage.stsci.edu/2005/12a

26 April 2005

SDSS uses 200,000 Quasars to Confirm Einstein's Prediction of Cosmic Magnification

The presence of dark matter and giant clusters of galaxies are predicted to have an effect on the brightness of distant quasars when their light is bent through gravitational lensing by applying a subtle magnification component. The prediction, made using Einstein's General Theory of Relativity, has been hard to confirm for the past two decades because of the lack of a large enough sample of quasars with accurate brightness measurements, and those results that were obtained were hard to match with current theoretical cosmological models.

New data from the Sloan Digital Sky Survey has provided a huge sample of 200,000 quasars, and provided astronomers with a new measurement of the subtle magnification effect that matches the predictions of light from the far off quasars as it passes through regions of dark matter and galaxies.

"Observing the magnification effect is an important confirmation of a basic prediction of Einstein's theory," said Bob Nichol of the University of Portsmouth, England. "It also gives us a crucial consistency check on the standard model developed to explain the interplay of galaxies, galaxy clusters and dark matter."

The advantage this team had was the digital database from Sloan, and some new computer techniques to sift through the database to detect quasars. The sample was ten times larger than possible with conventional detection methods.

"Now that we've demonstrated that we can make a reliable measurement of cosmic magnification, the next step will be to use it as a tool to study the interaction between galaxies, dark matter, and light in much greater detail," said Andrew Connolly of the University of Pittsburgh.

http://www.sdss.org/news/releases/20050426.magnification.html

http://xxx.lanl.gov/abs/astro-ph/0504510

Above: *This sharpest-ever image of the Whirlpool Galaxy (M51) and a companion galaxy, taken with the Advanced Camera for Surveys aboard NASA's Hubble Space Telescope, illustrates a spiral galaxy's grand design, from its curving spiral arms, where young stars reside, to its yellowish central core, home of older stars. Image courtesy NASA, ESA, S. Beckwith (STScI) and The Hubble Heritage Team (STScI/AURA).*

30 April 2005

Astronomers Confirm the First Image of a Planet Outside Our Solar System

The first image of a planet outside our solar system taken by the European Southern Observatory's (ESO) Very Large Telescope (VLT) has been confirmed by astronomers, ending a year-long discussion about the object. New images convinced astronomers that the red object circling a brown dwarf was a planet weighing in about five times the mass of Jupiter.

"Our new images show convincingly that this really is a planet, the first planet that has ever been imaged outside of our solar system," says Gael Chauvin of ESO.

The planet, called 2M1207b, lies 200 light years from Earth in the constellation of Hydra. The planet shines one hundred times fainter and lies 55 astronomical units away from its parent star. The spectrum of the object revealed water molecules, indicating it was a relatively cold object and not a small brown dwarf, but a bona-fide planet.

"The two objects - the giant planet and the young brown dwarf – are moving together; we have observed them for a year, and the new images essentially confirm our 2004 finding," says Benjamin Zuckerman of UCLA. "I'm more than 99% confident."

http://www.eso.org/outreach/press-rel/pr-2005/pr-12-05.html

http://xxx.lanl.gov/abs/astro-ph/0504659

Above: *An artist's impression of the first planet outside of our solar system to be imaged orbiting a brown dwarf (right). The giant planet is approximately five times the mass of Jupiter. Both objects are believed to be very young, less than 10 million years old. The brown dwarf is still surrounded by a circumstellar disc. Image courtesy ESO.*

MAY 2005

10 May 2005

NASA's Chandra Observatory Catches X-Ray Super-Flares

The Chandra Orion Ultradeep Project (COUP) announces new results that imply our Sun was very active with huge X-ray flares during its early stages of formation. In the deepest X-ray observations of a star cluster ever made, the orbiting Chandra X-Ray Observatory focused on the Orion Nebula for 13 days, producing a wealth of data. This unprecedented look into the heart this famous star formation region allows scientists to study the X-ray behavior of young sun-like stars with ages between 1 and 10 million years.

The results imply that super-flares torched our young solar system, flares that dwarf, in energy, size and frequency, anything seen from our Sun today. Some theories of planet formation suggest that such superflares create turbulence in the planet forming accretion disk, preventing the small proto-planets from migrating inwards. If such ideas are correct, we have the counterintuitive concept that such massive flares actually enhanced the survival chances of the proto-Earth (see **http://chandra.harvard.edu/photo/2005/orion/najita.html**.) The COUP data generated the most uniform and comprehensive dataset on the X-ray emission of normal stars ever obtained in the history of X-ray astronomy.

http://chandra.harvard.edu/photo/2005/orion/

http://lanl.arxiv.org/abs/astro-ph/0506650

http://www.astro.psu.edu/coup/

11 May 2005

Swift Catches Unique Short Gamma-Ray Burst

Within the first few months of operation, the Swift orbiting gamma-ray observatory detects the precise location of a very short GRB, lasting only 50 milliseconds. Prior to Swift, there was almost no data on such short bursts. The precise location revealed its proximity to an elliptical galaxy. This detection provided the first observational support to the theory that short busts are produced from different processes than longer bursts. Short bursts are thought to be the product of the catastrophic merger of two neutron stars, or a neutron star and a black hole. The massive collapse of a single star in a hypernova, in contrast, produce longer-duration bursts.

"Seeing the afterglow from a short gamma-ray burst was a major goal for Swift, and we hit it just a few months after launch," said Dr. Neil Gehrels. He is the Swift project scientist at NASA's Goddard Space Flight Center (GSFC), Greenbelt, Md. "For the first time, we have real data to figure out what these things are," he added. (See the special feature article about GRBs and the Swift observatory written by Neil Gehrels and Peter Leonard elsewhere in this volume.)

http://swift.gsfc.nasa.gov

http://grb.sonoma.edu

Above: *This deep Chandra image of the Orion Nebula Cluster shows over 1,600 X-ray sources. Most are young stars in the Nebula and the rest are either background galaxies or foreground stars. The red X-ray sources are mostly young stars with little absorption by intervening gas and the blue X-ray sources are mostly young stars with larger amounts of gas absorption. Image courtesy NASA/CXC/Penn State Univ./E.Feigelson and K.Getman et al.*

12 May 2005

First-Ever Infrared Flash Challenges Old Notion of Nature's Biggest Bang

Nature's brightest blasts also produce a flash in the infrared. A robotic telescope, responding to the detection of a gamma-ray burst (GRB), helped detect the first, and unexpected, infrared signature from these huge explosions. The Peters Automated Infrared Imaging Telescope (PAIRITEL) in Arizona found the burst just six minutes after Europe's gamma-ray observatory, Integral, detected the GRB, and followed it for three days until it became too faint. The Keck observatory provided quick follow-up infrared observations to confirm the discovery by PAIRITEL. (Swift also caught the burst, its 3rd detection, during commissioning phase of the new observatory).

Gamma Ray Bursts such as this one are thought to be created when a massive star collapses to form a black hole. Jets, containing relativistic winds, form near the black hole, and strike shells of material surrounding the black hole, generating shock waves and x-ray and optical afterglows. If conditions are just right, an infrared afterglow can be produced. Detecting such glows allow scientists to probe the immediate surroundings of the star that led to the GRB.

http://www.cfa.harvard.edu/press/pr0513.html

http://pairitel.org

http://integral.esac.esa.int/

15 May 2005

Astrophysicists Propose a Novel Procedure to Probe the Accelerating Expansion of the Universe and the Nature of Dark Energy

The discovery that the expansion rate of our universe is accelerating came is a shock to astronomers in the late 1990s. "The accelerating expansion of the universe constitutes one of the most intriguing and challenging problems in astrophysics", says Dr. Mustapha Ishak-Boushaki, a research associate at Princeton University in New Jersey.

Theorists have introduced the idea of "Dark Energy" that has a negative (repulsive) gravitational force to drive the acceleration. The amount required is a staggering two-thirds of the entire energy density of the universe. The big question is, does dark energy suggest a whole new form of energy, or is it the observational signature of a failure of Einstein's General Theory of Relativity? In either case, the ramifications are profound.

Professor David Spergel, of Princeton University, claims "Our goal is to be able to distinguish the two cases." Ishak-Boushaki and Spergel have proposed a new technique that might differentiate between the two possibilities. The technique involves identifying certain signatures from a modified theory of gravity based on multiple dimensions (as suggested by String theory). If the acceleration is due to Dark Energy then the expansion history of the universe should be consistent with the rate at which clusters of galaxies grow. Deviations from this consistency would be a signature of the breakdown of General Relativity at very large scales of the universe. The proposed procedure implements this idea by comparing the constraints obtained on Dark Energy from different cosmological probes and allows one to clearly identify any inconsistencies.

As an example, a universe described by a 5-dimensional modified gravity theory was considered in this study and it was shown that the procedure can identify the signature of this theory. Importantly, it was shown that future astronomical experiments can distinguish between modified gravity theories and Dark Energy models.

http://www.utdallas.edu/physics/faculty/profiles/ishak.html

http://www.aas.org/publications/baas/v37n4/aas207/dcblock.html

15 May 2005

Canada's Space Telescope Discovers Star Ringing Out of Tune

The Canadian suitcase-sized space telescope, MOST (Microvariability & Oscillations of STars), is producing a steady stream of results. Three of them are:

1. The vibrations of a nearby sun-like star called Eta Boötis are out of tune compared to the predictions of theoretical models. Eta Boötis is the third brightest star in the constellation of Boötes, the Herdsman. Dr. Guenther of Saint Mary's University, in Halifax, Nova Scotia comments, "Our understanding of the Sun and other stars is only as good as our ability to match data to models. Better data from new instruments like

MOST forces us to produce better models." The computer simulations are now so sophisticated that they take six months to compute.

2. A planet orbiting the star HD209458a, discovered by Spitzer Space Telescope in 2005, has provided some clues about its atmosphere to MOST. By watching for the dip in light when the planet disappears behind the star, scientists claim that the planet is less reflective than a planet like Jupiter. The way the planet reflects light back to us is sensitive to atmospheric composition and temperature. MOST Mission Scientist Dr. Jaymie Matthews claims "This is telling us about the nature of this exoplanet's atmosphere, and even whether it has clouds." The planet appears to be reflecting 30-40% of starlight whereas Jupiter reflects about 50%.

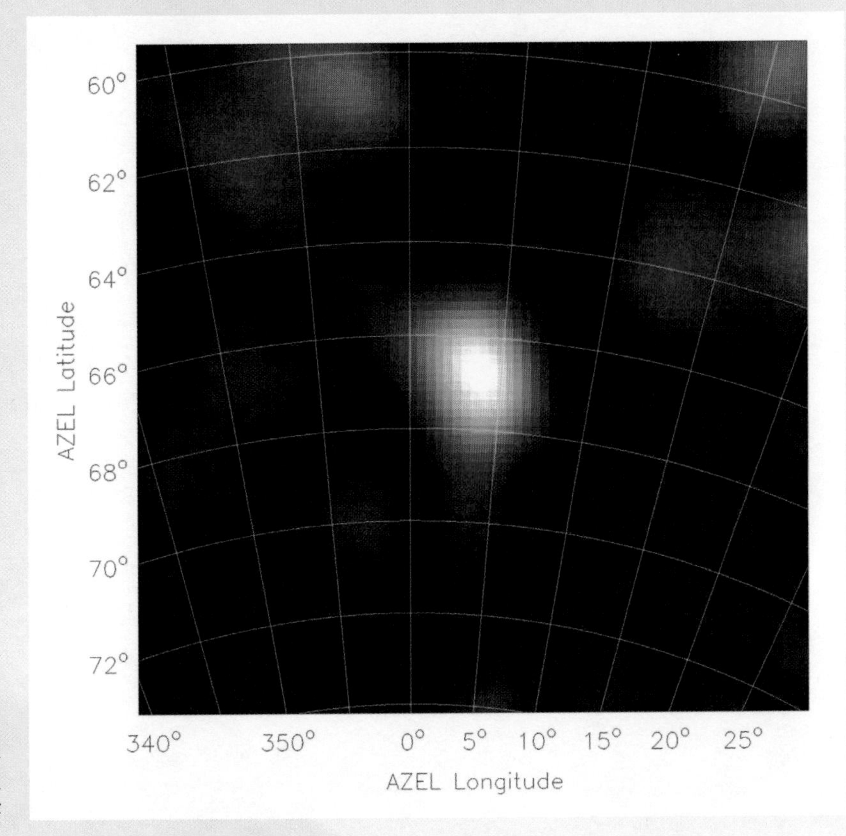

3. The star, Tau Bootis, known to have a planet since its discovery in 1997, has now shown by MOST to exhibit subtle variations in the light output that are in synchronization with the orbit of the planet. The best explanation put forward is that the planet, orbiting at 1/20th the Earth-Sun distance, is forcing the outer envelope of the star to rotate so that it keeps the same face to the planet. "The interactions between the star and the giant planet in the Tau Bootis system are unlike anything astronomers have seen before," commented Dr. Matthews. The unusual aspect of this, since captured rotation is not unusual between planets and their moons, is the fact that the planet is probably under 1% of the star's mass. It is likely that only the surface layers of the star are affected.

http://www.astro.ubc.ca/MOST/

http://www.astro.ubc.ca/MOST/science.html#results

18 May 2005
Ultra-High Energy Cosmic Particles Produce Brilliant Radio Flash

Bizarre flashes of radio light, a thousand times brighter than the Sun and almost a million times faster than normal lightning, lasting up to 30 nanoseconds, have been detected using a prototype of the new high-tech radio telescope. The startling flashes span an area of sky twice that of the full Moon.

The LOPES (LOFAR Prototype Experimental Station) is a test-bed for the European-wide LOFAR radio telescope, designed, among other things, to detect ultra high energy cosmic rays. The origin of these particles, with energies above 10^{10} eV, remains a mystery. Ultra high energy cosmic rays strike our atmosphere, causing a cascade of elementary particles descending through the atmosphere. Some of these daughter products can reach the ground and be detected by instruments like LOPES.

New technology incorporated into LOPES, combined with other planned telescopes like Pierre Auger and the Square Kilometer Array promise to create a cascade of data to unravel the mystery of these extreme cosmic rays. The expected rate of the brightest flashes is about one per day.

http://www.mpifr-bonn.mpg.de/public/pr/pr_lopes.html

http://www.astro.ru.nl/lopes/

■ ***Above:*** *The image shows a false-color radio map of a part of the low-frequency radio sky above the LOPES experiment at the time of a cosmic ray hit. The bright blob in the center of the image is the radio flash which lit up for about 30 nanoseconds. Image taken from Falcke et al. 2005, Nature, 19 May 2005 issue, 'Detection and localization of atmospheric radio flashes from cosmic ray air showers', and reproduced courtesy of Heino Falcke and the LOPES collaboration.*

23 May 2005

Astronomers, Amateur Skywatchers Find New Planet 15,000 Light Years Away

A new planet in a solar system roughly 15,000 light years from Earth has been detected using the gravitational microlensing technique. The Optical Gravitational Lensing Experiment, or OGLE, noticed that the star in question was starting to move in front of another star that was even farther away, near the center of our galaxy. Over a period of time the closest star's gravitational influence on starlight focuses light from the more distant star, thus creating a distinctive and unique brightening signature. However, a month later, when the more distant star had brightened a hundred-fold, astronomers detected a new and unexpected pattern in the signal -- a rapid distortion of the brightening -- that could only mean one thing – the existence of an orbiting planet. This is the second planet detected using microlensing. The first one, found in 2004, is estimated to be at a similar distance.

Two amateur astronomers in New Zealand using 10- and 14-inch telescopes, coordinated by the Microlensing follow-up Network, helped adding to the global 24 hour coverage of data collection.

The planet weighs in about three times the mass of Jupiter, and lies about three astronomical units from its parent star. The technique that astronomers used to find the planet worked so well that the team thinks it could be used to find much smaller planets -Earth-sized planets, even very distant ones.

http://www.cfa.harvard.edu/press/pr0514.html

http://arXiv.org/abs/astro-ph/0505451

http://bulge.princeton.edu/~ogle/

Above: *This artist's impression shows the second planet to be found using gravitational microlensing. It weighs about three times as much as Jupiter and orbits a sun-like star located approximately 15,000 light-years from the Earth. Image courtesy David A. Aguilar (Harvard-Smithsonian Center for Astrophysics).*

26 May 2005

Core Collapse in Naked Carbon/ Oxygen Stars May Be Source of Gamma-Ray Bursts

One possible candidate for certain gamma-ray bursts is the Type Ic supernova, also known as a hypernova. The theory of how these stars collapse, called the collapsar model, predicts certain tell-tale signatures within the spectral fingerprint long after the supernova has erupted. New evidence of SN 2003jd obtained with the Keck 10-m and Subaru 8-m telescopes on Hawaii, while not absolute, does provide strong evidence that the model is correct.

The idea involves an asymmetrical explosion of a naked carbon/oxygen star, flattening as it collapses, and producing a bipolar beam of matter and energy that generates the intense gamma- ray burst. The iron core of the star collapses to a black hole, and the increased rotation rate forces the infalling matter into a disk, preventing release of matter along the equator, and forcing two blowholes to appear near the poles. The lack of a hydrogen or helium envelope results in an increased chance that the jet will punch out of the supernova. If the jet is along our line of site, we see a gamma-ray burst. If the jet is not along our line of site and no gamma-ray burst is seen, then some time after the explosion, we should see emission lines of oxygen and other heavy elements, but split into pairs. The reason for this is that with the jets oriented away from our line of sight, opposite sides of the expanding equatorial disk are seen, one side is moving towards us, and the opposite side is moving away. Such opposing motions create a Doppler shift in the spectral lines and creating pairs.

The observations of SN 2003jd, lying 260 million light years away, are the first time such pairs of spectral lines have been observed in type 1c supernovae. Kawabata, Mazzali and his team analyzed the spectra, revealing that they exhibit split oxygen and magnesium emission lines exactly as would be expected if the collapsar model of gamma-ray production were correct.

http://arxiv.org/abs/astro-ph/0505199

30 May 2005

Spitzer Captures Fruits of Massive Stars' Labors

A new Spitzer Space Telescope image casts a whole new light on the famous massive star, Eta Carinae, and associated nebula. Violent stellar winds from Eta Carinae have sent shock waves through the giant cloud of interstellar material, and astronomers expected new stars to be forming within it. The new infrared image from Spitzer, which pierces the normally opaque dusty nebula, has indeed revealed vast numbers of embryonic stars.

"We knew that stars were forming in this region before, but Spitzer has shown us that the whole environment is swarming with embryonic stars of an unprecedented multitude of different masses and ages," said Dr. Robert Gehrz, from the University of Minnesota.

Huge finger-like pillars of dust point away from Eta Carinae, and the image

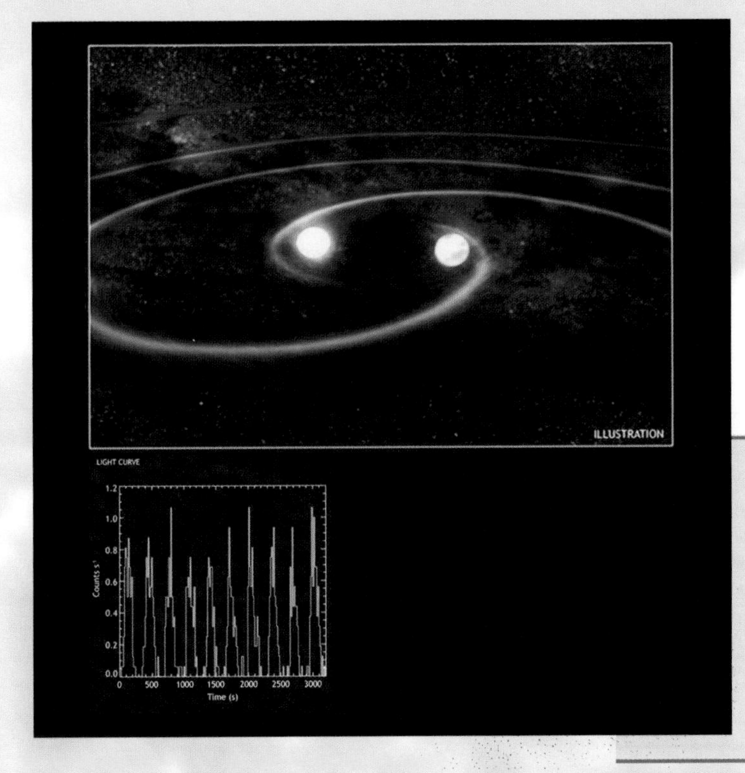

■ ***Left:*** *The graph compiled from Chandra observations of RX J0806.3+1527 (J0806) show that its X-ray intensity varies with a period of 321.5 seconds. This implies that J0806 is a binary star system where two white dwarf stars are orbiting each other (as shown in the adjacent illustration), approximately every 5 minutes. Light curve courtesy NASA/CXC/GSFC/Tod Strohmayer. Image courtesy NASA/ GSFC/Dana Berry. Story on page 17.*

reveals the stellar swarms buried within them. At 10,000 light years from Earth, Eta Carinae weighs in at 100 times the mass of the Sun. It is so unstable that astronomers expect it to explode as a supernova in the next few hundred years.

The stars within the nebula come in a wide range of ages and mass. The 200 light year-wide Carina Nebula provides a vast experimental region to help us understand how just one gas cloud can produce so many stars.

http://www.spitzer.caltech.edu/Media/releases/ssc2005-12/release.shtml

Above: *This false-color image from NASA's Spitzer Space Telescope shows part of the Carina Nebula. The infrared telescope penetrates this dusty cloud to reveal the heat from warm, embedded star embryos (yellow or white) tucked inside deeper, more buried finger-like pillars of thick dust (pink). Hot gases are green and foreground stars are blue. In the inset visible-light picture of the Carina Nebula, dust pillars are fewer and appear dark because the dust is soaking up visible light. Main image courtesy NASA/JPL-Caltech/N. Smith (Univ. of Colorado at Boulder). Inset courtesy National Optical Astronomy Observatory.*

30 May 2005

Andromeda Galaxy Three Times Bigger in Diameter Than Previously Thought

Astronomers using the new Keck/DEIMOS multi-object spectrograph with the 10m Keck II telescope have determined that the Andromeda Galaxy is three times larger than previously thought. The result comes from careful measurements of faint stars previously thought to be in a giant halo around the famous M31 galaxy. Instead, the spectrographic measurements revealed these stars to be orbiting the galaxy in a vast extended disk.

The new measurements indicate the diameter of Andromeda Galaxy is 220,000 light years. Previous measurements indicated M31 spanned 70,000-80,000 light years across.

The new measurements come from about 3,000 stars too faint to detect with spectrographs on smaller telescopes. Earlier studies indicated these stars were part of a chaotic halo that arose as smaller galaxies crashed into Andromeda. The stars in that case would be going in different directions. Instead, the measurements showed the stars were part of an organized disk extending far beyond the typical disk seen in images of M31.

Spectrographic measurements reveal the radial velocity of stars towards or away from the observer. Spectral lines shift to longer wavelengths if a star is moving away and to shorter wavelengths if it is moving towards us. If stars are in a disk, objects on one side of the galaxy will appear to be moving opposite to those on the other side, and can be assumed to be orbiting the center of the galaxy, as observed in this situation.

http://www.keckobservatory.org/news/science/050530.html

http://arxiv.org/abs/astro-ph/0504164

30 May 2005
Chandra Sees Orbiting Stars Flooding Space with Gravitational Waves

NASA's Chandra X-ray Observatory has found evidence that two white dwarf stars are orbiting each other in a death grip so strong, they're destined to merge. The system may be the strongest source of gravitational waves yet known. Known as RX J0806.3+1527 or J0806, the pair of white dwarfs are only 80,000 kilometers apart.

Optical and x-ray observations show periodic variations of 321.5 seconds. Astronomers believe this to be the orbital period of the pair, though concede that it could represent the spin rate of one of the white dwarfs. "If confirmed, J0806 could be one of the brightest sources of gravitational waves in our galaxy," said Tod Strohmayer of NASA's Goddard Space Flight Center, Greenbelt, Md.

Einstein's General Theory of Relativity predicts a binary star system should emit gravitational waves that rush away at the speed of light. This loss of energy results in an orbital period decrease as the stars move closer together. The Chandra X-ray observatory measured the orbital period decay of J0806 at 1.2 milliseconds every year – in line with relativity predictions. This means the stars are moving closer together at about 60 centimeters per day.

White dwarfs are very compact stars that have used up all their fuel. They are often found in the centers of planetary nebulae. They pack about half the mass of the Sun into a sphere about the size of Earth.

http://chandra.harvard.edu/press/05_releases/press_053005.html

http://arxiv.org/abs/astro-ph/0504150

JUNE 2005

6 June 2005
Cornell astronomers find key evidence supporting theory of quasars

Active galactic nuclei come in a variety of types. The prevailing idea is that the different types are the result of viewing the same type of object from different angles, namely a central black hole with an accretion disk embedded in a torus of dust, and bipolar jets. Observational evidence of the tell-tale infrared absorption features of silica dust in type 2 AGNs should result in corresponding emission features in type 1 AGNs, which until now remained unconfirmed

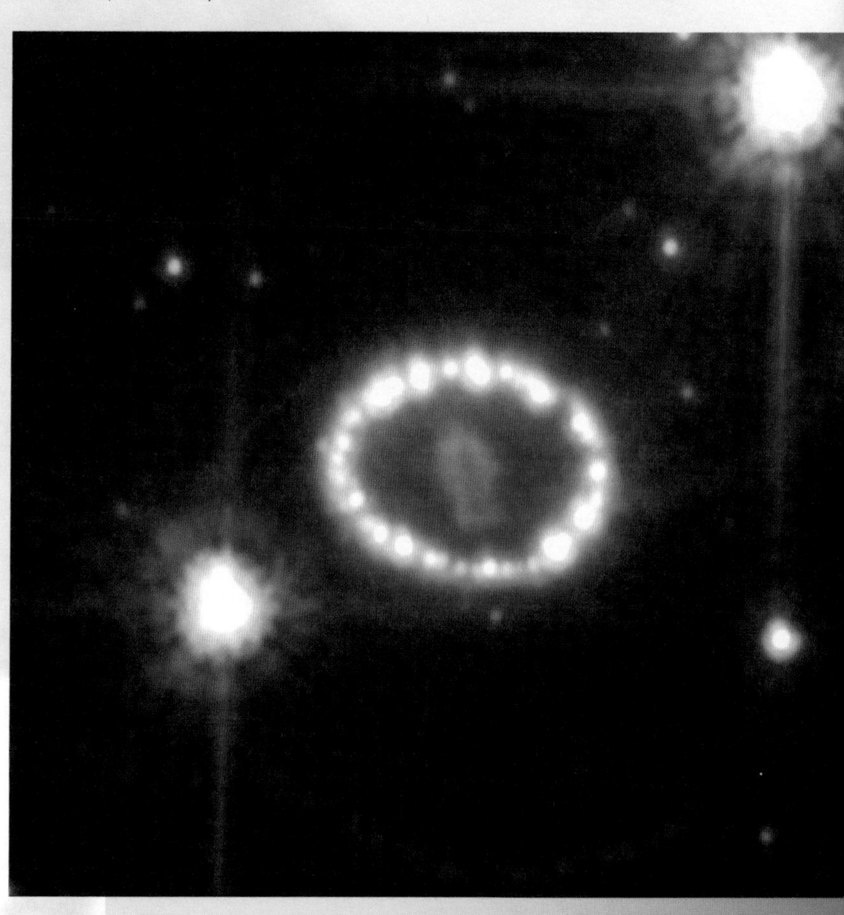

■ **Above:** *The remnant of supernova 1987A shows no sign of the neutron star scientists believe is lurking at its heart (see story on page 19). The Hubble Space Telescope took this image in December 2004. Image courtesy P. Challis and R. Kirshner (Harvard-Smithsonian Center for Astrophysics).*

for years. New observations using the Spitzer Space Telescope's infrared spectrograph show the elusive emission features at 10 and 18 microns in a study of five quasars by astronomers from Cornell University and other institutions. Type 1 AGNs are viewed 'face-on' and the new sensitivity of the Spitzer Space Telescope allowed the detection to be made.

http://www.news.cornell.edu/stories/June05/spitzer.quasars.hao.lg.html

Above: *This false-color picture of the supernova remnant Cassiopeia A comprises images taken by three of NASA's Great Observatories, using three different wavebands of light. Infrared data from the Spitzer Space Telescope are colored red; visible data from the Hubble Space Telescope are yellow; and X-ray data from the Chandra X-ray Observatory are green and blue. While Spitzer reveals warm dust in the outer shell at temperatures of a few hundred degrees Kelvin, Hubble sees the delicate filamentary structures of hot gases about 10,000 degrees Kelvin, and Chandra probes incredibly hot gases, up to about 10 million degrees Kelvin. Chandra can also see Cassiopeia A's neutron star (turquoise dot at center of shell). Image courtesy NASA/JPL-Caltech/O. Krause (Steward Observatory).*

6 June 2005

Spitzer Sees the Infrared Echo of Neutron Star Outburst

The 325-year old Cassiopeia A supernova remnant had a surprise in store for astronomers studying the speed of ejecta from the old explosion. Some of the fast-moving wisps and knots were in the wrong place to have been ejected during the original explosion. Instead, they appear to be the result of energy ejected by the neutron star about 50 years ago.

Using NASA's Spitzer Space Telescope, Oliver Krause and George H. Rieke of the University of Arizona's Steward Observatory took exposures a year apart and compared them. The brightest features appear to be moving close to the speed of light, suggesting a light echo from a recent outburst is traveling outwards, and illuminating clouds of material. "We see light that is just now encountering two interstellar clouds and heating them so we see the outburst's infrared echo," Rieke said.

The observations suggest the central neutron star shot tremendous energy into the surrounding nebula in opposite directions. Because the directions of both jets are oriented away from our line of sight, the original outburst would have gone unnoticed, even if astronomers were looking at the precise moment of the outburst.

http://www.spitzer.caltech.edu/Media/releases/ssc2005-14/index.shtml

http://mips.as.arizona.edu/mipspage/

http://xxx.lanl.gov/abs/astro-ph/0506186

6 June 2005
Exploding Star Left No Visible Core

The puzzling mystery of the lack of a central neutron star in the SN1987A remnant continues nearly 20 years after the dramatic supernova illuminated one end of the Large Magellanic Cloud. A new report announces that the Hubble Space Telescope has been unable to detect anything of the central compact object. The observations raise questions about why the neutron star is not seen.

"Therein lies the mystery -- where is that missing neutron star?" mused co-author Robert Kirshner of the Harvard-Smithsonian Center for Astrophysics (CfA).

The original mass of the star that exploded was about 20 times that of the Sun, right on the borderline where it is difficult to predict the outcome of the central compact object.

Here are the top candidates:

Black Hole – detectable only if material is being accreted and falling into the black hole, heating up as it does so and emitting copious amounts of radiation before disappearing across the event horizon.

Pulsar – theory predicts these objects are formed between 100 and 1,000 years after the initial explosion. A neutron star could be there, but the intense magnetic field required to be a pulsar has not yet developed. Such strong magnetic fields are required for the pulsar beam that would illuminate surrounding gas clouds.

A violent and chaotic-looking mass of gas and dust is seen in this Hubble Space Telescope image of a nearby supernova remnant. Denoted N 63A, the object is the remains of a massive star that exploded, spewing its gaseous layers out into an already turbulent region. Image courtesy NASA, ESA, HEIC, and the Hubble Heritage Team (STScI/AURA), with thanks to Y.-H. Chu and R. M. Williams (UIUC). *(see story on page 20, June 7.)*

All observations to date have failed to detect a light source in the center of the supernova remnant, leaving the question of the outcome unanswered.

http://www.cfa.harvard.edu/press/pr0515.html

http://arxiv.org/abs/astro-ph?0505066

7 June 2005
Supernova Remnant Menagerie

A young supernova remnant, N 63A, shocks the surrounding gas clouds creating a spectacular and chaotic mass of gas and dust. A new image released from the Hubble Space Telescope reveals this supernova remnant, a member of the star-forming region N 63, located in the Large Magellanic Cloud (LMC).

As with all Hubble images, details reveal far more than just an attractive image. The colors represent light emitted by oxygen (shown in blue), hydrogen (shown in green) and sulphur (shown in red). Narrow-band filters isolate these colors and allow detailed analysis of the structure of the remnant. Very massive stars lie in the immediate vicinity of N 63A and the 'mother-star', or progenitor, of the supernova that produced the remnant seen here was about 50 times more massive than our own Sun. The supernova exploded within a giant cavity surrounding the original star that was produced by the strong stellar winds that such a massive star produces during its lifetime. When the expanding supernova shock reaches the edge of the slower moving gas bubble, spectacular celestial fireworks result. (Editor: The best example of such fireworks are currently being observed in another LMC supernova, SN 1987A, which is described in a feature article elsewhere in this volume, and in an image accompanying the 6 June story of 1987A.)

http://www.spacetelescope.org/news/html/heic0507.html

http://arxiv.org/abs/astro-ph/0209370

13 June 2005
Astronomers Announce the Most Earth-Like Planet Yet Found Outside the Solar System

The race to be the first to find an Earth-like planet continues unabated, and each month or two, new claims are announced. This latest discovery of a planet about 7.5 times as massive as the Earth is a significant step in attempting to detect smaller and smaller planets. The planet, potentially the first rocky object ever found, orbits the star Gliese 876 just 15 light years from Earth. Located in the constellation of Aquarius, the planet orbits the star in a mere two days.

However, there's no chance for life on this planet – it's short period places it close to the star, resulting in surface temperatures between 400 to 750 degrees Fahrenheit (200 to 400 degrees Celsius).

All of the nearly 150 other extrasolar planets discovered to date around normal stars have been larger than Uranus, an ice-giant about 15 times the mass of the Earth.

http://www.keckobservatory.org/news/science/gl876/050613.html

15 June 2005
Ripples in Cosmic Neutrino Background Measured for First Time

The first evidence of ripples in the universe's primordial sea of neutrinos has been reported. While impossible to measure directly, scientists combined data from WMAP and the Sloan Digital Sky Survey to arrive at the result. The findings confirm predictions of both Big Bang theory and the Standard Model of particle physics.

Neutrinos are elementary particles with no charge and very little mass, and are extremely difficult to study due to their very weak interaction with matter. According to the standard Big Bang model, neutrinos permeate the universe at a density of about 150 per cubic centimeter. The Earth is therefore immersed in an ocean of neutrinos without us ever noticing.

Dr. Roberto Trotta of the University of Oxford, England, said: "This research provides important new evidence in favor of the current cosmological model, unifying it with fundamental physics theories. Cosmology is becoming a more and more powerful laboratory where physics not easily accessible on Earth can be tested and verified. The high quality of recent cosmological data allows us to investigate neutrinos in the cosmological framework, obtaining measurements which are competitive with -- if not superior to -- particle accelerator findings."

15 June 2005

SMA Stares Into the Throat of a Cosmic Jet

Located 1000 light years away, and less than a thousand years old, in the constellation of Perseus, lies the Herbig-Haro object, HH 211. Known to be young stars undergoing formation, buried inside dark clouds of dust and gas, many Herbig-Haro objects have been found to sport energetic jets that burst through the dense clouds, revealing themselves to the outside universe.

Now under the scrutiny of the Submillimeter Array (SMA), detail of the region close to where the jet forms in HH 211 has been obtained, providing valuable data to confirm new models of how jets arise. Submillimeter wavelengths are able to penetrate the thick clouds surrounding Herbig-Haro objects.

How these jets form and how they influence planet formation is of tremendous interest to astronomers. A major step to their understanding

■ ***Above:*** *Herbig-Haro 211 consists of two jets of material, visible at lower right, blasting from a young protostar hidden behind dust. The Sub-Millimeter Array (SMA) has looked deep within the inner regions of the jets, close to their launching point, in order to test predictions of jet formation models. This infrared image was taken using the FLAMINGOS camera, which was designed and constructed at the University of Florida. Image courtesy A.A. Muench-Nasrallah and Harvard-Smithsonian Center for Astrophysics.*

came from a model constructed by astronomer Hsien Shang of the Academia Sinica Institute of Astronomy and Astrophysics (ASIAA) and her colleagues.

They created a model of jet formation that calculates temperatures, densities and brightnesses within stellar jets, and compared the model to real observations from SMA. The challenge is that the model can predict what happens so close to the surface of the star, near where the jet originates (8 million kilometers above the star's surface), that observations have been unable to confirm the predictions until now.

"Our model predicts what we will see about 100 astronomical units from the star," Shang said, "(and) we can begin to look at the HH 211 system at the scale of the model and test those predictions. So far, everything checks out."

Reaching finer resolution, closer to the surface of the star and the origin of the jet, will require a larger array of submillimeter telescopes such as the Atacama Large Millimeter array (ALMA) being constructed in Chile. It's due to come on line in 2012.

HH 211, discovered in 1994, is so young that it is still growing by accreting matter from the dense surroundings of gas and dust. The buried protostar will eventually be similar in mass to our Sun.

http://www.cfa.harvard.edu/press/pr0518.html

http://xxx.lanl.gov/abs/astro-ph/0512252

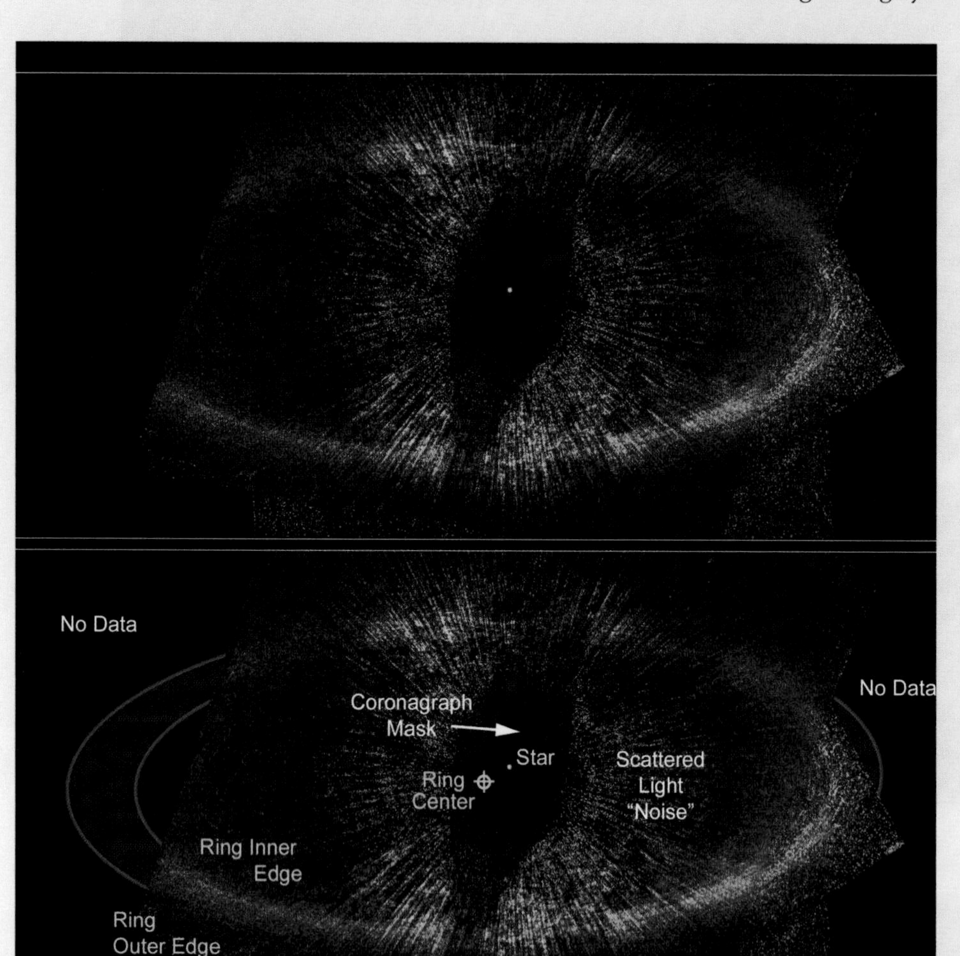

Above: *The upper panel shows the first visible-light image of a dusty ring around the nearby, bright young star Fomalhaut. Astronomers used the Advanced Camera for Surveys' (ACS) coronagraph aboard the Hubble Space Telescope to block out the light from the bright star so they could see the faint ring. The left part of the ring is outside the telescope's view. The dot near the ring's center marks the star's location. Astronomers believe that an unseen planet moving in an elliptical orbit is reshaping the ring. The lower view points out important features in the image. Image courtesy NASA, ESA, P. Kalas and J. Graham (University of California, Berkeley), and M. Clampin (NASA's Goddard Space Flight Center)*

22 June 2005
NASA's Hubble Chases Unruly Planet

The bright star, Fomalhaut, has given up more secrets about its circumstellar disk of dust. The Hubble Space Telescope took new images of the dust belt and found evidence of an unseen planet gravitationally tugging on the disk.

Fomalhaut, a star only 25 light years distant and about twice the mass of the Sun, is a mere 200 million years old - one-twentieth the age of the sun. In our solar system, that age was the epoch of bombardment, when asteroids and comets rained down on the planets and moons. However, the dust ring is ten times older than those seen around AU Microscopium and Beta Pictoris.

The center of the ring around Fomalhaut has a clear offset, amounting to 2.3 billion kilometers, from the star and provides strong evidence for at least one unseen planetary mass object orbiting it.

"Our new images confirm those earlier hypotheses that proposed a planet was perturbing the ring," said astronomer Paul Kalas of the University of California at Berkeley. "The ring is similar to our solar system's Kuiper Belt, a vast reservoir of icy material left over from the formation of our solar system planets."

In addition, the ring's inner edge is sharper than its outer edge. A dark object must be ploughing this region clear, since an unperturbed dust ring would not have a clear boundary. A third feature is the ring's relatively narrow width, about 3.7 billion kilometers. Without an object to gravitationally keep the ring material intact, the particles would spread out much wider.

http://hubblesite.org/news/2005/10

http://arxiv.org/abs/astro-ph/0506574

23 June 2005

'Bumpy Space Dust' Explains Origin of Most Common Molecule in Universe

Where did all that molecular hydrogen come from? This question has raised a mystery about why hydrogen, the most abundant element in the universe, is largely in molecular form and not as single atoms. An answer has come from an unexpected source. A trivial feature of dust grains could be the culprit.

Ohio State University researchers led by Dr Eric Herbst recently found that the solution could be a result of whether the surfaces of interstellar dust grains are smooth or bumpy.

"For two hydrogen atoms to have enough energy to bond in the cold reaches of space, they first have to meet on a surface," explained Eric Herbst.

His group modeled the motion of two hydrogen atoms tumbling along the different surfaces until they found one another to form a molecule. They found the formation of molecular hydrogen in the right proportions only occurred on bumpy surfaces. The new models will be used to simulate other chemical reactions in space.

Below: *An artist's impression of the Giant Magellan Telescope in its enclosure. The completed GMT will include seven 8.4-meter diameter mirrors, making a huge telescope with a resolving power equivalent to that of a single 24.5-meter mirror. Image courtesy Giant Magellan Telescope - Carnegie Observatories. Artwork by Todd Mason, Mason Productions.*

27 June 2005

Giant Magellan Telescope Begins Mirror Fabrication

The University of Arizona's Steward Observatory Mirror Lab is pre-firing its huge spinning furnace and inspecting tons of glass for casting a first 8.4-meter (27-foot) diameter mirror for the Giant Magellan Telescope (GMT). The casting is scheduled for Saturday, 23 July. With this milestone step, the GMT becomes the first extremely large ground-based telescope to start construction. The GMT will include seven 8.4-meter mirrors, making a huge telescope with the resolving power equivalent to that of a 24.5 meter diameter mirror. Its completion is expected to be 2016, and produce images ten times sharper than the Hubble Space Telescope.

http://mirrorlab.as.arizona.edu/

http://www.gmto.org

29 June 2005

Scientist Refines Cosmic Clock To Determine Age of Milky Way

By comparing the decay of two long-lived radioactive elements, uranium-238 and thorium-232, Nicolas Dauphas of the University of Chicago has determined the age of the Milky Way at approximately 14.5 billion years, plus or minus more than 2 billion years.

That age generally agrees with the estimate of 12.2 billion years -nearly as old as the universe itself - as determined by methods using both globular clusters and white dwarfs – methods that rely heavily on stellar evolution and nuclear physics.

This new method is free of previous assumptions. The uranium/thorium ratio in a single old star that resides in the halo of the Milky Way is measured by spectroscopy. What

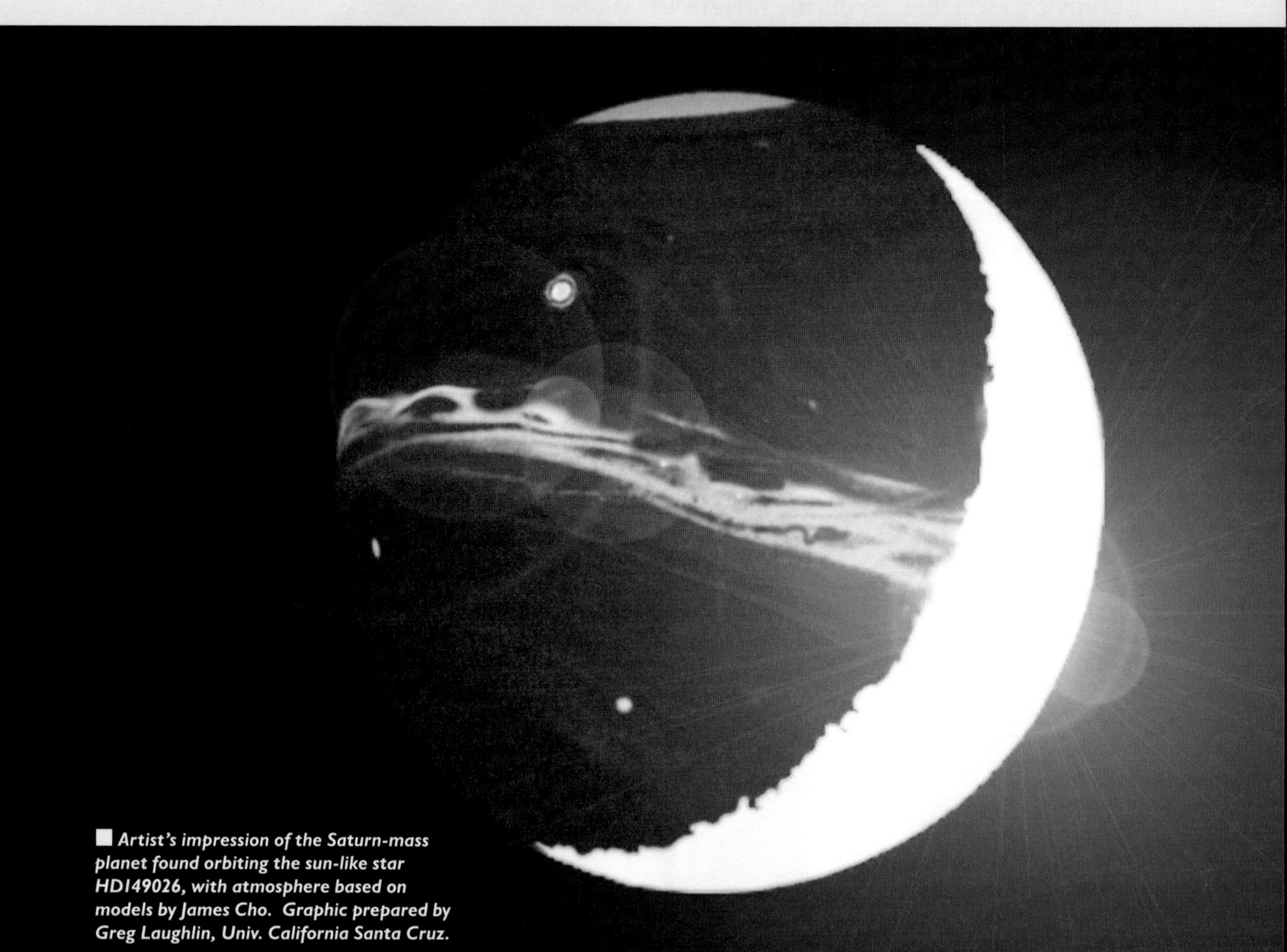

Artist's impression of the Saturn-mass planet found orbiting the sun-like star HD149026, with atmosphere based on models by James Cho. Graphic prepared by Greg Laughlin, Univ. California Santa Cruz.

is not known is the original ratio when the star formed. However, Dauphas' new technique combines the known decay rate of uranium-238 and thorium-232, and the amount remaining in meteorites that are 4.5 billion years old, which include debris from earlier generations of stars with different U-238/Th-232 ratios, to determine the all-important starting value for the uranium/thorium clock.

"Age determinations are crucial to a fundamental understanding of the universe," said Thomas Rauscher, an assistant professor of physics and astronomy at the University of Basel in Switzerland. "The wide range of implications is what makes Nicolas' work so exciting and important." Scientists can now use that clock to determine the age of a variety of interstellar objects and particles, including cosmic rays.

http://www-news.uchicago.edu/releases/05/050629.milkyway.shtml

30 June 2005

Astronomers Using the VLT Discover Bright Cosmic Mirage Far Away

Using ESO's Very Large Telescope, European astronomers have discovered an amazing cosmic mirage, known to scientists as an Einstein Ring. This cosmic mirage, dubbed FOR J0332-3557, is seen towards the southern constellation Fornax (the Furnace), and is remarkable on at least two counts. First, it is a bright, almost complete Einstein ring. Second, its distance of 7 billion light years makes it the farthest ever found.

Gravitational lenses occur when a foreground galaxy and the background galaxy are perfectly lined up. Light from the background galaxy is bent by the gravity of the nearby galaxy, distorting the original galaxy's image into a ring. It's called an Einstein ring because Albert Einstein's theory of General Relativity predicted their occurence, as first described by Chwolson and Link in the early twentieth century.

Because the amount of lensing is determined by the mass of the intervening galaxy, the method provides a powerful tool. The distant galaxy's mass is estimated at one trillion suns. In addition, it allows unseen distant galaxies to be magnified. The distant galaxy lies 12 billion light years from Earth, and existed at a time when the universe was 12% of its present age.

http://www.eso.org/outreach/press-rel/pr-2005/phot-20-05.html

30 June 2005

Astronomers Discover Planet with Largest Solid Core

The planet, orbiting the sun-like star HD 149026, is roughly equal in mass to Saturn, but it is significantly smaller in diameter. Luckily for astronomers, the planet in question passes in front of the star. "When that happens, we are able to calculate the physical size of the planet, whether it has a solid core, and even what its atmosphere is like," said Debra Fischer. She is consortium team leader and professor of astronomy at San Francisco State University, Calif.

It takes just 2.87 days to circle its star. By developing a sophisticated model of the planet's structure, theorists show it has a solid core approximately 70 times Earth's mass.

"For theorists, the discovery of a planet with such a large core is as important as the discovery of the first extrasolar planet around the star 51 Pegasi in 1995," said Shigeru Ida, theorist from the Tokyo Institute of Technology, Japan.

Scientists have rarely had opportunities like this to collect such solid evidence about planet formation. More than 150 extrasolar planets have been discovered by observing changes in the speed of a star as it moves toward and away from Earth. Fortunately this planet also happens to pass right on front of the star, allowing the careful measurements to be made.

The deductions lend support for the "core accretion" model for planetary formation. Scientists have two competing but viable theories about planet formation. With the "core accretion" theory, planets start as small rock-ice cores that grow as they gravitationally acquire additional mass. Scientists believe the large, rocky core of this planet could not have formed by the competing theory of "gravitational instability", where planets form during a rapid collapse of a dense cloud. They think it must have grown a core first, and then acquired gas.

http://tauceti.sfsu.edu/n2k/

http://tauceti.sfsu.edu/n2k/satotransit.pdf

JULY 2005

11 July 2005

A New Survey - SDSS-II Will Map the Universe, the Milky Way and Dark Energy

Funding for continuing the most ambitious astronomical sky survey ever has been completed, allowing the Sloan Digital Sky survey to continue through 2008 and enter a new phase of its work. SDSS-II will complete observations of a huge contiguous region of the northern skies and will study the structure and origins of the Milky Way galaxy and the nature of dark energy.

In late June 2005 the funding package for a new three-year venture called the Sloan Digital Sky Survey II (SDSS-II) was completed, led by the Alfred P. Sloan Foundation of New York City, the National Science Foundation (NSF), the U.S. Department of Energy and the member institutions.

SDSS-II has three components; LEGACY, SEGUE and the new Sloan Supernova survey.

LEGACY will complete the SDSS survey of the extragalactic universe, obtaining images and distances of nearly a million galaxies and quasars over a continuous swath of sky in the Northern Hemisphere.

SEGUE, or the Sloan Extension for Galactic Understanding and Exploration, will map the structure and stellar makeup of the Milky Way galaxy, and gathering data on how the Milky Way formed and evolved.

"The SEGUE project will allow us for the first time to get a 'big picture' of the structure of our own Milky Way," explained consortium member Heidi Newberg of Rensselaer Polytechnic Institute.

Below: *The Sloan Digital Sky Survey (SDSS) uses a dedicated, 2.5-meter telescope on Apache Point, New Mexico, equipped with two powerful special-purpose instruments. The 120-megapixel camera can image 1.5 square degrees of sky at a time, about eight times the area of the full moon. A pair of spectrographs fed by optical fibers can measure spectra of (and hence distances to) more than 600 galaxies and quasars in a single observation. A custom-designed set of software pipelines keeps pace with the enormous data flow from the telescope. Image courtesy Fermilab Visual Media Services.*

"Identifying the oldest stars will help us understand how the elements of the periodic table were formed long ago inside of stars," Newberg said.

A new intensive study of supernovae will allow astronomers to precisely measure the distances of distant supernovae, using them to map the rate of expansion of the universe. "This study will help to verify and quantify one of the most important discoveries of modern science - the existence of the cosmological dark energy," explained consortium member Andy Becker of the University of Washington.

http://www.sdss.org/

12 July 2005

X-Ray Oscillations From Biggest Star Quake In Universe Provide Clues to Mysterious Interior of Neutron Stars

On 27 December 2004, a gigantic explosion on a neutron star halfway across the Milky Way galaxy, the largest such explosion ever recorded in the universe, triggered rapid vibrations that were detected by the Rossi X-ray Timing Explorer. The explosion was so bright that it lit the Earth's upper atmosphere and blinded all the X-ray satellites in space for an instant.

"This explosion was akin to hitting the neutron star with a gigantic hammer, causing it to ring like a bell," said Richard Rothschild, an astrophysicist at the University of California's Center for Astrophysics and Space Sciences.

The quakes ripped through the neutron star at an incredible speed, vibrating the star at 94.5 cycles per second. "This is near the frequency of the 22nd key of a piano, F sharp," said Tomaso Belloni, an Italian member of the team who measured the signals.

The peculiar oscillations the researchers found began three minutes after a titanic explosion on a neutron star that, for only a tenth of a second, released more energy than the sun emits in 150,000 years. The oscillations then gradually receded after about 10 minutes.

SGR 1806-20 is the formal designation of the neutron star that exploded and sent X-rays flooding through the galaxy.

Much as geologists probe the Earth's interior from seismic waves produced by earthquakes, this neutron star quake should allow astronomers

■ ***Above:** Artist's impression of the 27 December 2004 gamma-ray flare from the neutron star SGR 1806-20 impacting Earth's upper atmosphere. Image courtesy NASA.*

to probe the interiors of these mysterious stellar objects and provide critical information about the internal structure of neutron stars.

http://ucsdnews.ucsd.edu/newsrel/science/mcgamma.asp

http://www.esa.int/science

http://arxiv.org/abs/astro-ph/0505255

13 July 05
First Planet Under Three Suns Is Discovered

An extrasolar planet orbiting a triple star system has been discovered in the constellation Cygnus, 149 light years from Earth, using the 10-meter Keck I telescope in Hawaii. The planet appears to be slightly larger than Jupiter. Maciej Konacki, a senior postdoctoral scholar in planetary science at Caltech, discovered the Jupiter-class planet orbiting the main star of the close-triple-star system known as HD 188753. The three stars are about as close to one another as the distance between the Sun and Saturn.

Konacki refers to the new type of planet as "Tatooine planets," because of the similarity to Luke Skywalker's view of his home planet's sky in the first Star Wars movie.

If the planet's surface was solid, the view would be strange indeed. The planet orbits the star in a mere three and a half days. The parent star is similar to the Sun, yellow in color. The other two stars are orange and red.

"The environment in which this planet exists is quite spectacular," says Konacki. "With three suns, the sky view must be out of this world-literally and figuratively."

The formation of so called "Hot Jupiters" so close to their parent star still puzzles scientists. Now that an example has been found in a multiple stars system, the theories of planet formation will be put under pressure to come up with ideas of how such a planet could form in such a complicated setting.

http://www.keckobservatory.org/news/050713.html

http://xxx.lanl.gov/abs/astro-ph/0509490

http://xxx.lanl.gov/abs/astro-ph/0509767

http://xxx.lanl.gov/abs/astro-ph/0511463

21 July 2005
This Supernova Just Won't Fade Away

M100 is a favorite galaxy of both amateur and professional observers. It's a bright face-on spiral galaxy located 56 million light years away in the constellation of Coma Berenices. In 1979, a supernova (SN 1979C) went off in the outer spiral arms of M100 and faded quickly.

However, recent observations by the orbiting European XMM-Newton X-ray observatory shows that the supernova is as bright in x-rays as it was when it was discovered.

"This 25-year-old candle in the night has allowed us to study aspects of a star explosion never before seen in such detail," said Dr Stefan Immler, leader of the team, from NASA's Goddard Space Flight Center, USA.

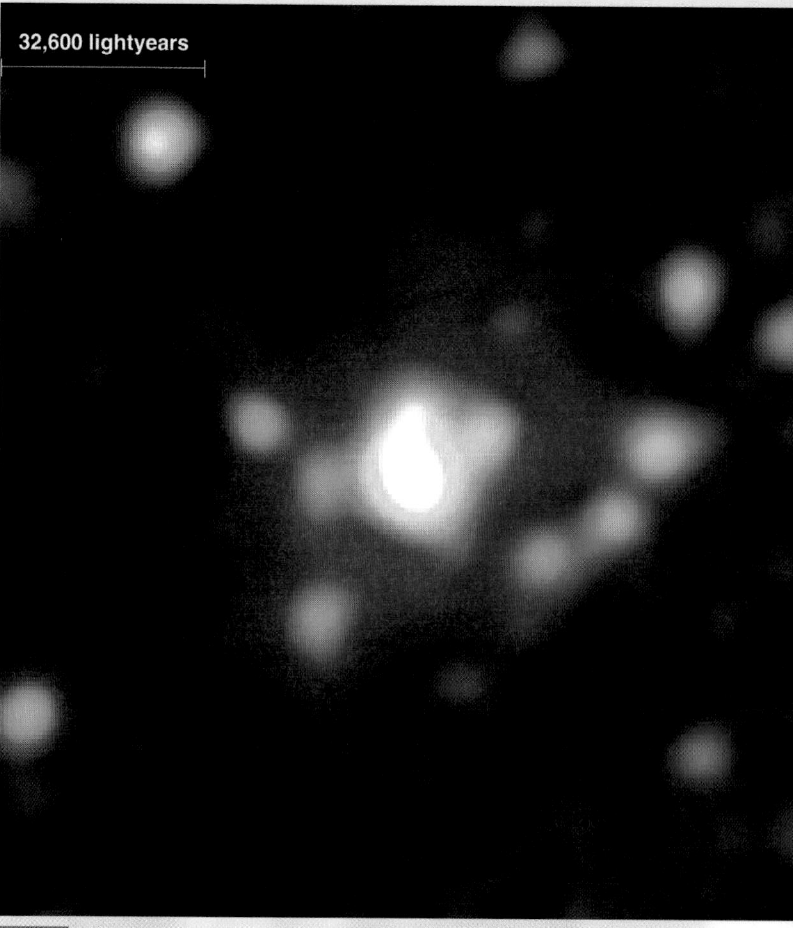

Right: *This is an XMM-Newton image of X-ray light from the galaxy M100. Instead of the prominent spiral arms, apparent in optical images, the X-ray image reveals high-energy activity throughout the galaxy, particularly at the centre. The red and orange regions are sources of very hot, diffuse gas between stars. The supernova SN 1979C is the orange hotspot about 7 o'clock from the white galactic centre. Image courtesy ESA/NASA/Immler et al.*

Astronomers found that the original star that exploded was about 18 times more massive than our Sun. Such stars produce fierce stellar winds for thousands of years. Now the outflowing shock wave from the supernova is striking this previously ejected material, producing the X-rays. Twenty five years after the supernova, the shock wave has reached material ejected in stellar winds a long time ago. It's effectively illuminating the past 16,000 years' worth of stellar wind activity.

"We can use the X-ray light from SN 1979C as a 'time machine' to study the life of a dead star long before it exploded," said Immler.

Ultraviolet observations from XMM-Newton confirm the existence of a vast reservoir of material previously ejected from the star in its hot stellar wind. It covers a region 25 times the size of our solar system and is a thousand times denser than the solar wind – a more modest outpouring of material from our own star.

http://www.esa.int/esaSC/SEME2C0DU8E_index_0.html

http://arxiv.org/abs/astro-ph/0503678)

25 July 2005

Extreme Galactic Disk: A Rare Peek into Galaxy's Past

A nearby spiral galaxy, NGC 4625, is revealing some of its secrets to NASA's orbiting Galaxy Evolution Explorer (GALEX) observatory. Imaging galaxies in the ultraviolet, GALEX reveals the hottest and most active regions of recent star formation within galaxies. The surprising find in NGC 4625 is a set of young spiral arms that extend far beyond the normal spiral arms seen in visible light.

"We are excited about this particular galaxy," commented Armando Gil de Paz, one of the Carnegie Observatories team that undertook the observations with GALEX. "Not only is it the most extended UV galaxy disk found so far, the galaxy appears to be in its most active star-formation stage. Surprisingly, it has properties typical of galaxies found farther away, when the universe was much younger."

The new spiral disk extends 28,000 light years from the galaxy center, a staggering four times the optical radius. The hot blue stars in this new disk may have formed from the inflow of fresh gas and dust from interaction with another nearby galaxy, NGC 4618. NCG 4618 is peculiar since it has no spiral arms – they appear to have been stripped, leading to the speculation that it had something to do with the formation of the giant arms in NGC 4625. Another newly discovered galaxy, NGC 4625A, may also be involved. NGC 4625 is about 31 million light-years away in the constellation Canes Venatici.

"It's a spectacular opportunity to see the conditions in which spiral galaxies like our own Milky Way formed," remarked Samuel Boissier, of the Carnegie Observatories. "This galaxy will provide a fundamental test for the models of galaxy formation."

http://www.galex.caltech.edu/

http://xxx.lanl.gov/abs/astro-ph/0506357

Above: *This image highlights the hidden spiral arms (blue) that were discovered around the nearby galaxy NGC 4625 (top) by the ultraviolet eyes of NASA's Galaxy Evolution Explorer. Although very faint in visible light, the spiral arms are bright in ultraviolet, because they are bustling with hot, newborn stars that radiate primarily ultraviolet light. The youthful arms stretch out to a distance four times the size of the galaxy's core. Image courtesy NASA/JPL-Caltech/Carnegie Observatories/DSS*

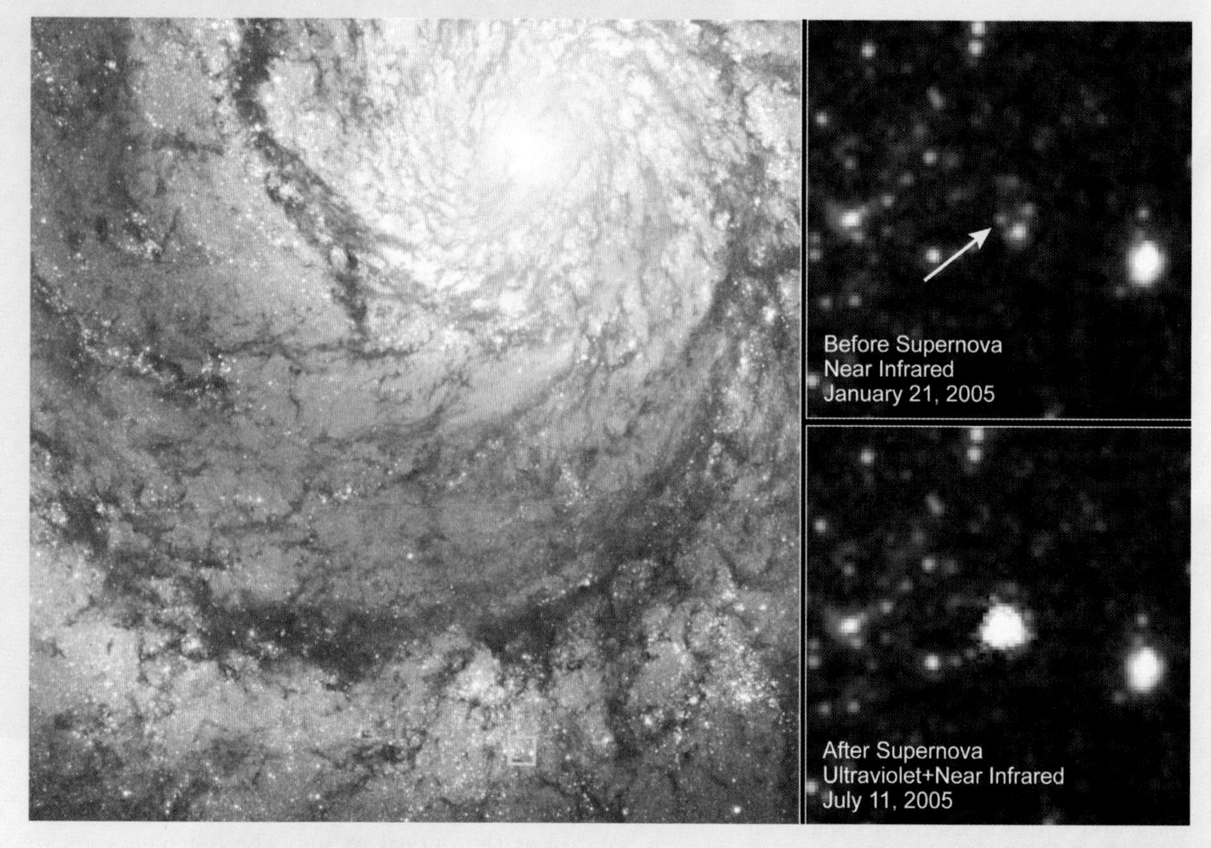

28 July 2005

Spitzer Finds Life Components in Young Universe

Molecules of wide variety have been found throughout space using radio telescopes for decades. Many molecules also emit discrete wavelengths of light in the infrared part of the spectrum. NASA's Spitzer Space Telescope has now found some signatures of complex organic molecules within galaxies at a tremendous distance, the farthest yet known. The discovery has some significant implications, because the distance of the galaxies place them at a time when the universe was one quarter of its current 13.7 billion year age.

Above: *Hubble pinpoints doomed star that explodes as supernova. The color image at left shows a section of the Whirlpool Galaxy (M51) taken in January 2005 with the Advanced Camera for Surveys aboard NASA's Hubble Space Telescope. The small green square marks where the supernova's progenitor star resides. The lower-right image shows a picture of SN 2005cs (the central bright object), taken 11 July 2005, by Hubble. By comparing the lower-right image with the color image at left, astronomers identified the supernova's progenitor star [marked by the arrow in the (pre-explosion) upper-right image]. The star was found to be a red supergiant whose mass is seven to 10 times that of the Sun. Image courtesy NASA, ESA, W. Li and A. Filippenko (University of California, Berkeley), S. Beckwith (STScI), and The Hubble Heritage Team (STScI/AURA).*

"This is 10 billion years further back in time than we've seen (the molecules) before," said Dr. Lin Yan of the Spitzer Science Center at the California Institute of Technology

The spectral fingerprint is from molecules called polycyclic aromatic hydrocarbons (PAH's). These complex molecules are made up of atoms of carbon and hydrogen, and are considered by most scientists to be among the building blocks for proteins, and consequently, life.

Two of the eight sources observed by Spitzer are in very distant starburst galaxies, undergoing rapid bursts of star formation. While very faint optically due to plenty of obscuring dust, the same dust glows brightly in the infrared.

"These complex compounds tell us that by the time we see these galaxies, several generations of stars have already been formed," said Dr. George Helou of the Spitzer Science Center. "Planets and life had very early opportunities to emerge in the Universe."

http://www.spitzer.caltech.edu/Media/releases/ssc2005-15/index.shtml

http://xxx.lanl.gov/abs/astro-ph/0504336

28 July 2005

Hubble Pinpoints Doomed Star in M51 That Exploded As Supernova

The spectacular galaxy, M51, has been a target of the Hubble Space Telescope ever since it was launched. In the spring of 2005 a massive mosaic image of the galaxy was released to the public. As luck would have it, in late June, a bright supernova erupted in one of the spiral arms. It was even bright enough to be detected in amateur telescopes. Now, Hubble has pinpointed the original star that exploded.

Twelve days after the supernova appeared, astronomers took new images of the galaxy and compared those to the earlier images to identify the progenitor, and its identity was confirmed by the Canada-France-Hawaii telescope on Mauna Kea, Hawaii.

The supernova was a Type II event, a class of supernova linked to the core collapse of a red giant star. The star that exploded was likely a red supergiant about nine times the mass of the Sun, on the lower end of the mass scale for a Type II supernova.

http://hubblesite.org/news/2005/21

http://xxx.lanl.gov/abs/astro-ph/0507502

http://xxx.lanl.gov/abs/astro-ph/0507394

Below: *This spectrum tracks mid-infrared light from an extremely luminous galaxy, located 10 billion light-years distant, at a time when the Universe was only 1/4 of its current age. The infrared spectrometer on the Spitzer Space Observatory identified characteristic fingerprints of complex organic molecules called polycyclic aromatic hydrocarbons, illustrated in the artist's concept in the inset. These large molecules comprised of carbon and hydrogen, are considered among the building blocks of life.*
Image courtesy NASA/JPL-Caltech/L. Yan (SSC/Caltech)

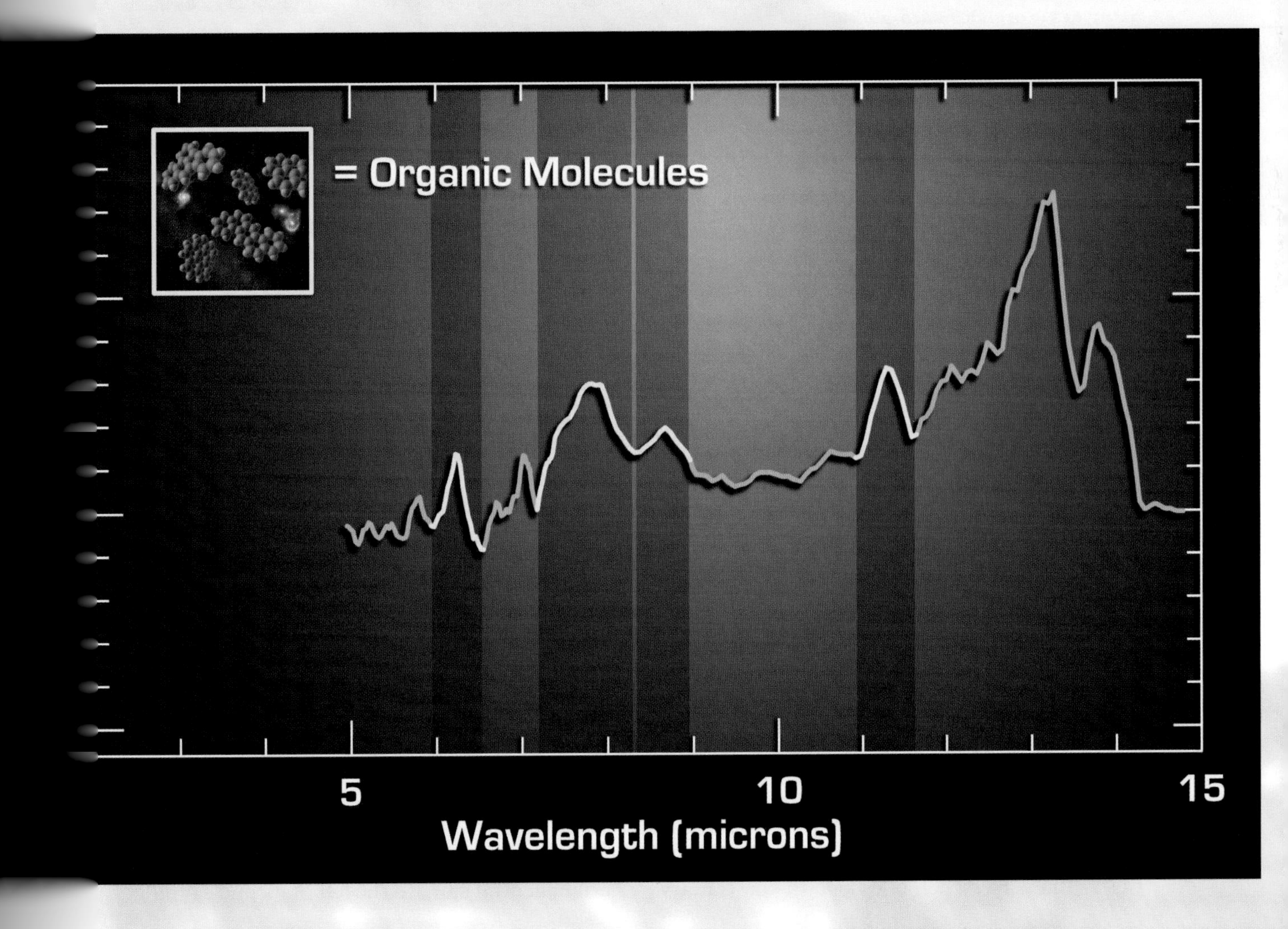

AUGUST 2005

1 August 2005

VLT Enables Most Accurate Distance Measurement to Spiral Galaxy NGC 300

The 16 long-period Cepheids observed by ESO's 8.2-m VLT Antu telescope have helped determine the most accurate distance yet to the spiral galaxy, NGC 300. The result is 6.1 million light years with an unprecedented uncertainty of 3%. The galaxy, similar to the Milky Way, is a prominent member of the Sculptor cluster of galaxies. The distance measurement is the first in the long-term Araucaria Project (**http://arxiv.org/abs/astro-ph/0509688**) designed to improve the local calibration of the distance scale with stellar standard candles such as Cepheids. A team of astronomers also used the Hubble Space Telescope for a similar distance determination.

http://www.eso.org/outreach/press-rel/pr-2005/pr-20-05.html

http://arxiv.org/abs/astro-ph/0405581.

See also http://arxiv.org/abs/astro-ph/0510298

3 August 2005

Spitzer Finds Hidden, Hungry Black Holes

Supermassive black holes are known to lie in the heart of quasars. They are surrounded by vast gas and dust rings. The black holes consume vast quantities of this matter, causing it to emit copious amounts of x-ray, ultraviolet and visible light. These huge objects make quasars the brightest objects in the universe.

By measuring the cosmic x-ray background, an estimate of the total number of quasars with supermassive black holes can be made. It turns out that the observed number falls far short of this. Astronomers have predicted that many of these behemoths are hidden behind walls of gas and dust, and that a space-borne infrared telescope would be required to spot them.

Now, NASA's Spitzer Space Telescope (SST) has met that task with resounding success. What SST found was the long-sought missing population of black holes located in quasars. Two teams issued their results. One group found twenty one new objects in a small patch of sky, and confirmed as quasars by the Very Large Array radio telescope in New Mexico and by the William Herschel Telescope in the Canary Islands. A second group found five hidden black holes with Spitzer among a group of 27 objects within an ultra-deep x-ray field of view.

"If you extrapolate our 21 quasars out to the rest of the sky, you get a whole lot of quasars," said Dr. Mark Lacy of the Spitzer Science Center, in Pasadena, California, and a member of one team. It appears that most quasars are positioned in just the right orientation for their dusty rings to hide their light. Yet while it blocks most wavelengths, infra red gets through, allowing their detection by Spitzer.

The second team from The University of Arizona agrees. "Five hidden black holes out of 27 is about what we should find according to the theoretical predictions," graduate student Jennifer Donley said, "but to know if the predictions are exactly right, we really need more sources. Fortunately, Spitzer is observing other fields with deep X-ray measurements, so we will soon be able to tell if we really understand the glowing X-ray sky."

http://www.spitzer.caltech.edu/Media/releases/ssc2005-17/release.shtml

http://arxiv.org/abs/astro-ph/0509124

4 August 2005

Hubble Spies a Zoo of Galaxies

A 40-hour image by the Hubble Space Telescope pointing at a small patch of sky in the constellation of Fornax has revealed a menagerie of galaxies. The deep image reveals hundreds of galaxies never seen before within a region of sky less than the area of the full moon.

Some galaxies are clearly fully formed. They tend to be the larger objects in the image, because they lie closer to us. Galaxies farther away are generally smaller in the new image, lying at such vast distances that their light has taken billions of years to reach us, and consequently their light left the galaxy when they were much younger. Seeing their image on the sky now is rather like looking at an old picture of your grandfather when he was a young boy.

About a dozen bright foreground stars that lie within our galaxy act as a veil through which we view the distant universe. Because of the presence of a four-vaned secondary mirror holder in the Hubble Telescope, all stars generate images with spikes, due to diffraction.

In large scale versions of the image, look for the eye-catching red galaxy to the lower left of the bright central star. To the left of its nuclear bulge

This composite image from the Hubble Space Telescope's Advanced Camera for Surveys took nearly 40 hours to complete. It reveals an enormous diversity of galaxy types within a small patch of sky a fraction of the area of the full moon. Among the thousands of galaxies in this image are several foreground stars (easily identified by their diffraction spikes). Many galaxies that appear small in this image are so distant that their light has taken billions of years to reach us. Image courtesy NASA, ESA, and The Hubble Heritage Team (STScI/AURA), with thanks to J. Blakeslee (JHU) and R. Thompson (University of Arizona).

is an odd looking arc. This arc is a gravitational-lensed distant galaxy. Light from the farther galaxy is bent around the nearby galaxy's nucleus to form this prominent distorted arc.

http://hubblesite.org/news/2005/20

http://arxiv.org/abs/astro-ph/0312511

http://arxiv.org/abs/astro-ph/0509641

16 August 2005

Galactic Survey Reveals a New Look for the Milky Way

Understanding the structure of our Milky Way galaxy from within is a hard job. Dust blocks our view of many stars, complicating the task. Yet a recent survey of 30 million stars in the galactic plane has provided a dramatic change to the view of our once neat spiral galaxy. For years astronomers have said our galaxy is similar in structure to M31 in Andromeda. This new survey dramatically changes this picture. It now appears we live in a barred spiral galaxy.

"This is the best evidence ever for this long central bar in our galaxy," says Ed Churchwell of the University of Wisconsin-Madison.

The GLIMPSE (Galactic Legacy Mid-Plane Survey Extraordinaire) survey is a comprehensive project to study stellar and dust content within the inner galaxy that began in December 2003. Astronomers have implied a bar in our galaxy in the past, but these new measurements show its length is 7,000 light years longer than earlier measurements – spanning 27,000 light years across the central bulge. In addition, the orientation of the bar places it at a 45° angle to our line of site to the center of the galaxy, making it significantly foreshortened.

The survey was performed using the Spitzer Space Telescope. Its sensitive infrared eye is able to penetrate the obscuring dust that is so prevalent in the plane of the galaxy. The results provide the most detailed view ever achieved of the inner regions of our galaxy.

http://www.news.wisc.edu/11405.html

http://www.astro.wisc.edu/sirtf/

http://xxx.lanl.gov/abs/astro-ph/0508325

17 August 2005
Supernova 1987A: Fast Forward to the Past.

Ever since Supernova 1987A appeared in the Large Magellanic Cloud in 1987, and evidence that a slower moving ring of material was standing in the way of a fast moving explosive cloud, astronomers have waited with bated breath to see the fireworks that were predicted to begin sometime early in the 21st century. The wait is certainly over.

The intense shock wave from the brightest supernova in 400 years is now striking the ring of material ejected from the star about 20,000 years before the star was ripped apart. Observations with the Chandra X-ray telescope show a dramatic increase in X-radiation. In addition, the Hubble Space Telescope has revealed a spectacular ring, lit up like a string of brilliant rubies. Dense knots of material are being blasted by the staggering shock wave that has taken nearly 20 years to cross the space in between.

*■ **Above:** These deep-field images of the faint surroundings of the Sculptor group galaxy NGC 300 from the Gemini Observatory revealed extremely faint stars in the galaxy's disk up to 47,000 light-years from the center – effectively doubling the previously known radius of the disk. (The small dark areas in the Gemini images are shadows from instrument "guide-probes".) Images courtesy Gemini Observatory*

Observations in previous years have shown individual knots brightening. Now the entire ring is illuminated, with X-ray observations revealing multimillion-degree gas representing the inner edge of a larger amount of material ejected a long time ago by the original blue supergiant star called Sanduleak -69° 202. Over the coming decades, the shock wave will pass through material ejected at earlier and earlier times by the progenitor star prior to its stellar explosion. As Richard McCray, one of the scientists involved in the Chandra research, explains, "Supernova 1987A will be illuminating its own past."

(Editor: See Richard McCray's feature article about SN1987A later in this volume.)

http://chandra.harvard.edu

http://xxx.lanl.gov/abs/astro-ph/0510442

http://xxx.lanl.gov/abs/astro-ph/0511355

18 August 2005
Swift Satellite Finds Newborn Black Holes

Nearly half of the longer gamma-ray bursts (GRB's) seen by the Swift satellite reveal a series of powerful "hiccups" after the initial blast. The gamma-ray burst called GRB 050502B that occurred on 2 May, 2005, provided new detail about this unexpected scenario. The initial burst lasted 17 seconds and was followed about 500 seconds later with a second X-ray spike about 100 times brighter than anything seen before. GRB's are proving far more complex than straight forward explosions.

"Stars are exploding two, three and sometimes four times in the first minutes following the initial explosion," said Prof. David Burrows of Penn State University. "First comes a blast of gamma rays followed by intense pulses of X-rays. The energies involved are much greater than anyone expected," he added.

There are a variety of theories to explain the phenomenon, but the most favored is that Swift is seeing a new black hole being formed. During this formation process, matter gets ejected in a jet that has dramatic effects on its surroundings. What may also be happening is that as jet of material shoots away from the dead star, some of it starts to fall back, generating tremendous shockwaves that emit the later X-ray bursts.

"The newly formed black hole immediately gets to work," said Prof. Peter Meszaros of Penn State, head of the Swift theory team. "We aren't clear on the details yet, but it appears to be messy. Matter is falling into the black hole, which releases a great amount of energy. Other matter gets blasted away from the black hole and flies out into the interstellar medium. This is by no means a smooth operation," he added.

"None of this was realized before simply because we couldn't get to the scene of the explosion fast enough," said Dr. Neil Gehrels of NASA Goddard Space Flight Center, Greenbelt, Md., Swift principal investigator.

(Editor: See Neil Gehrels and Peter Leonard's feature article about Swift and its GRB detections elsewhere in this volume.)

http://www.nasa.gov/vision/universe/watchtheskies/double_burst.html

http://xxx.lanl.gov/abs/astro-ph/0512615

26 August 2005
A Chinese Dragon and a Knotted Galactic Embrace

In a poetic combination of images, the Gemini Observatory reveals two contrasting regions of space. In a pair of images, one of a dusty region in our own galaxy, and another of two galaxies undergoing an interaction, the dynamic nature of the universe on different size and timescales are revealed.

A dark cold dust cloud called NGC 6559 resembles a ghostly dragon. The dense cloud of dust, about 7 light years across, blots out stars from within our Milky Way, revealing the fascinating structures in silhouette. It lies about 5,000 light years away within the constellation

■ ***Left:** According to a new survey, the Milky Way has a definitive bar feature – some 27,000 light-years in length – that distinguishes it from ordinary spiral galaxies, as shown in this artist's impression. The survey, conducted using NASA's Spitzer Space Telescope, sampled light from an estimated 30 million stars in the plane of the galaxy. Illustration courtesy NASA/JPL-Caltech/R. Hurt (SSC/Caltech), with thanks to Terry Devitt.*

of Sagittarius, within a degree of the famous Lagoon Nebula (M8).

On a completely different scale, spanning 150,000 light years, two galaxies are entwined in a galactic embrace. Located 100 million light years away in the direction of the constellation of Pisces, NGC 520 reveals a pair of galaxies that are possibly like the Andromeda galaxy and our own. Some astronomers speculate that such an image could be a glimpse of the fate that will befall our own Milky Way in about five billion years. Dozens of more distant galaxies are visible in larger displays of the image.

While both images show a static view, over long timescales changes would be apparent. Ian Robson, Director of the UK's Astronomy Technology Center, explains. "If we could see either of these objects as an extreme time-lapse movie made over millions of years, the galaxy pair would dance in a graceful orbital embrace that is likely similar to the fate between our Milky Way and the great Andromeda Galaxy, while the dusty cloud would probably resemble waving smoke from an extinguished candle."

http://www.gemini.edu/pio/

Above: *This image of the cool, dark dust cloud NGC 6559 was taken by the Gemini South telescope. The dense cloud of dust, about 7 light years across, is absorbing light from a more distant cloud of red ionised hydrogen and so looks black, revealing fascinating structures in silhouette. It lies about 5,000 light years away within the constellation of Sagittarius. Image courtesy Gemini Observatory.*

31 August 2005
How To Build a Massive Star

New observations of a young protostar, located 2,000 light years away in the constellation of Cepheus, are providing new information about how massive stars form. It has been suggested by some scientists that the larger mass objects form by the gradual assimilation of numerous already formed protostars within very crowded regions of space. The object is called HW2, a massive protostar within the Cepheus A star forming region. The discovery of a huge flattened disk of material containing as much as 8 times the mass of the Sun and extending out 30 billion miles from the young sun challenges the merger theory.

The observations were made with the Submillimeter Array in Hawaii. When combined with radio telescope observations that reveal a bipolar jet perpendicular to the disk of material, the evidence for gravitational accretion is strong. Merging protostars would not form such a tidy disk.

"In the past, theorists have had trouble modeling the formation of high-mass stars and there has been an ongoing debate between the merger versus the accretion scenarios." said astronomer Nimesh Patel of the Harvard-Smithsonian Center for Astrophysics (CfA). "We've found a clear example of an accretion disk around a high-mass protostar, which supports the latter while providing important observational constraints to the theoretical models."

http://www.cfa.harvard.edu/press/pr0527.html

http://xxx.lanl.gov/abs/astro-ph/0509637

New observations by the Submillimeter Array show that massive stars form by accretion, the same process that formed our Sun. A competing theory suggested that massive stars form when smaller protostars merge to form a single giant star. The discovery of a huge flattened disk of material extending nearly 30 billion kilometers from a massive protostar called HW2 challenges the merger theory. Illustration courtesy Christine Pulliam (Harvard-Smithsonian Center for Astrophysics).

SEPTEMBER 2005

12 September 2005

Most Distant Explosion in Universe Detected: Smashes Previous Record

A gamma-ray burst detected by the Swift satellite, and observed quickly by numerous telescopes around the world, has turned out to be the most distant burst ever detected to date. This powerful burst, called GRB 050904, signifies the death of a massive star and the birth of a black hole. The object has a redshift of 6.29, or a distance of about 13 billion light years from Earth. (The universe is thought to be 13.7 billion years old.) The previous most distant gamma-ray burst had a redshift of 4.5. GRB 050904 was also very long in duration, lasting over 200 seconds; most bursts last only about 10 seconds.

"This is uncharted territory," said Daniel Reichart of the University of North Carolina, who made the distance measurement using the SOAR (Southern Observatory for Astrophysical Research) telescope, located in Chile. "This burst smashes the old distance record by 500 million light-years. We are finally starting to see the remnants of some of the oldest objects in the universe."

Before Swift was launched, most astrophysicists thought that gamma-ray bursts would only be seen out to a redshift of 2 or 3, with their numbers declining beyond that distance. However, while still a graduate student at the University of Chicago, Dan Reichart produced a paper in 1999 with Don Lamb that showed bursts would be visible almost to the edge of the observable universe, out to 13.3 billion light years (a redshift of 20). These new observations show that their predictions may well be correct. The search is on for even more distant GRB's.

(Editor: See also news for 22 September.)

Below: *Gamma-ray bursts are the beacons of star death and black hole birth. This sequence of images shows the life cycle of a supermassive star. In such a star the end result of the nuclear fusion process, which keeps the star shining throughout its life, is a core of iron. At this stage, if the star is massive enough, the core collapses into a black hole. The black hole quickly forms jets, and shock waves reverberating through the star ultimately blow apart the outer shells. Illustration courtesy Nicolle Rager Fuller/NSF.*

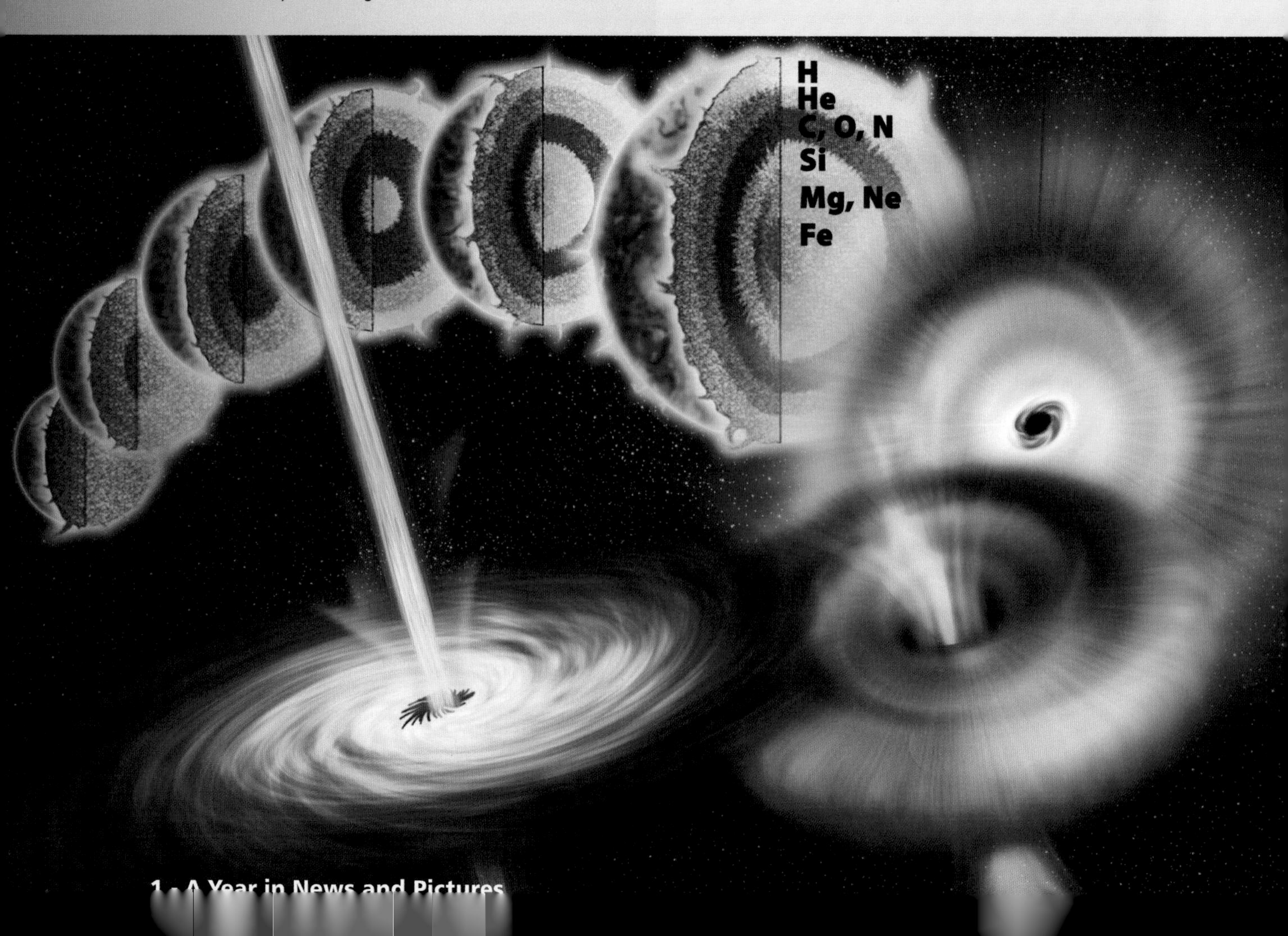

http://www.nasa.gov/vision/universe/starsgalaxies/2005_distant_grb.html

http://xxx.lanl.gov/abs/astro-ph/0509660

A copy of Lamb and Reichart's 2000 *Astrophysical Journal* paper is at

http://astro.uchicago.edu/home/web/lamb/lamb_reichart_2000.pdf.

20 September 2005
Hubble Finds Mysterious Disk of Blue Stars Around a Black Hole in M31

At the center of the Andromeda Galaxy (M31), astronomers have found a one light-year wide disk of blue stars, solving a decade-old mystery about the origin of the blue light at the galaxy's core. The stars are whizzing around the central massive core at amazing speeds - 3.6 million kilometers an hour (1,000 kilometers a second) – allowing a complete orbit in about 100 years.

The existence of a massive cluster of hot blue stars in such close proximity to a suspected central supermassive black hole offers a chance to measure the mass of the central object. The Space Telescope Imaging Spectrograph (STIS) was designed specifically to measure velocities of objects close in to suspected galactic black holes. When it targeted the core of M31, it measured speeds that require a mass of 140 million suns – three times larger than previously suspected. [Note: at its simplest, astronomers use the equation $F=mv^2/R$ to calculate the central mass]. These observations rule out the possibility that the dark central object is a massive collection of dead stars or neutron stars.

Ralf Bender of the Max Planck Institute for Extraterrestrial Physics in Garching, Germany comments, "And now we have proved that the black hole is at the center of a tiny disk of blue stars. But this uncovers a new mystery."

The new STIS observations show that the blue light consists of more than 400 stars and their properties suggest that they formed only 200 million years ago. Forming such a massive cluster of stars in the near proximity to a giant black hole presents problems that remains unsolved. The gravitational tidal forces on gas and dust are such that it makes it difficult for star formation to occur. (Editor: See also news for 13 October – one model suggests the mass of the disk could offset tidal forces).

"The blue stars are so short-lived that it is unlikely, in the 12-billion-year history of the galaxy, that a disk of them would appear just when we are ready to look for it," says team member Todd Lauer of the National Optical Astronomy Observatory. "That's why we think that the mechanism that formed this disk probably formed other stellar disks in the past and will form them again in the future."

However, astronomers are pleased with the new high resolution results. John Kormendy and a team of astronomers first discovered the central dark object in M31 in 1988, and strongly suspected it was a black hole.

"Right from the beginning, there were compelling reasons to believe that these are supermassive black holes," says Kormendy. "Looking for black holes always was a primary mission of Hubble. Nailing the black hole in Andromeda is an important part of its legacy. It makes us much more confident that the other central dark objects detected in galaxies are black holes too."

http://www.spacetelescope.org/news/html/heic0512.html

http://hubblesite.org/news/2005/26

http://xxx.lanl.gov/abs/astro-ph/0509839

22 September 2005
Ferreting Out the First Stars

The recent gamma-ray burst (see news for 12 September) with a redshift of 6.29 has astronomers thinking again about how far back the earliest GRB could be. While we have no observational data from the very first stars, and what they looked like, when they explode they are likely to be detectable. Consequently if a first generation star exploded as a hypernova, and emitted a gamma-ray burst, the Swift spacecraft should detect it.

The GRB 050904 burst erupted only a billion years after the Big Bang. Now, two theorists, Volker Bromm of the University of Texas at Austin and Avi Loeb of the Harvard-Smithsonian Center for Astrophysics, predict that one-tenth of the blasts Swift will spot during its operational lifetime will come from stars at a redshift of 5 or greater.

"Most of those GRBs will come from second generation or later stars," said Loeb. "But if we get lucky, Swift may even detect a burst from one of the very first stars that formed -- a star made of only hydrogen and helium." These so-

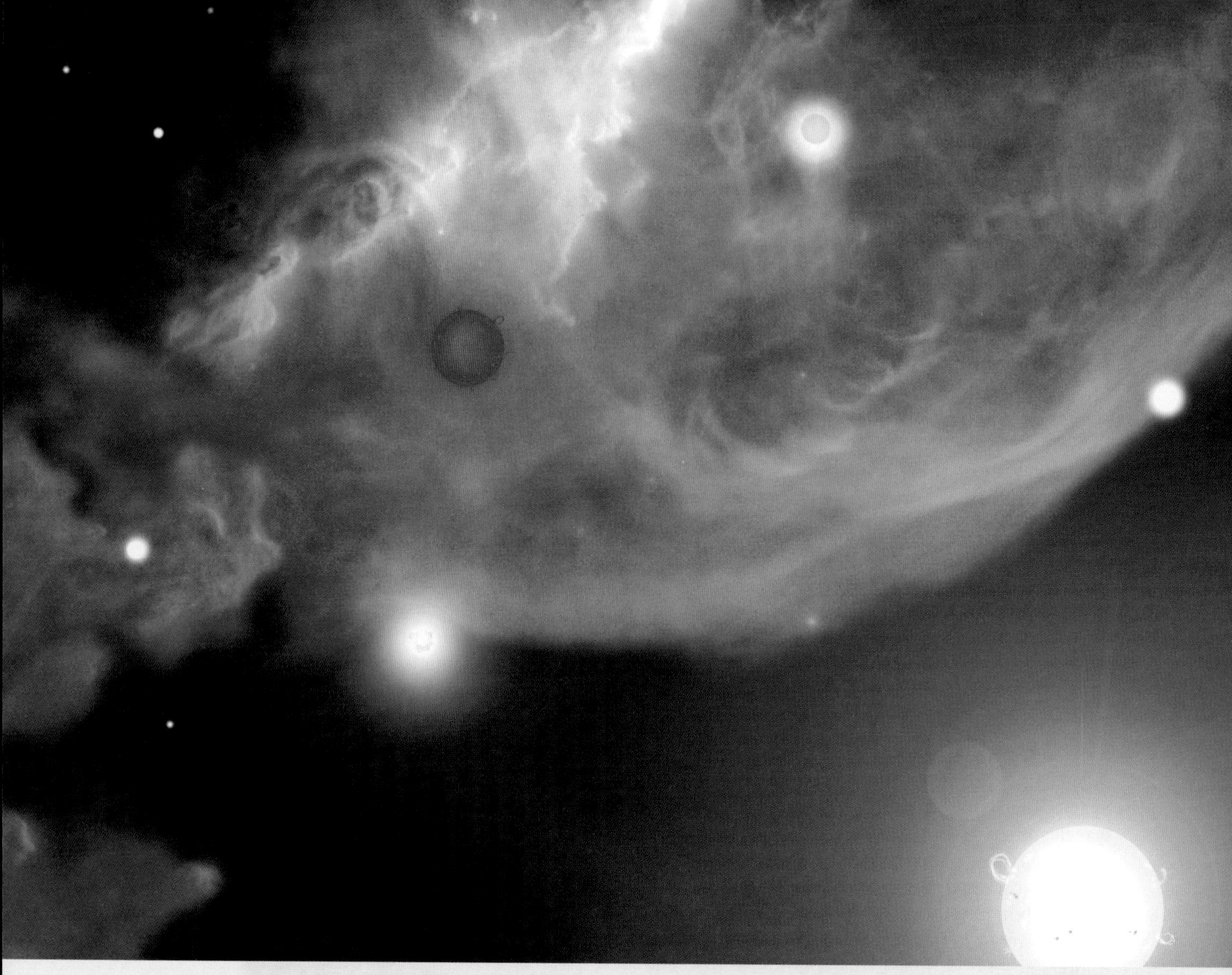

called Population III stars are the holy grail of stellar evolution. Some calculations suggest they could weigh in as much as 50-500 times the mass of our Sun. These huge stars would burn their fuel at very high rates and explode with such ferocity that their light would easily reach Earth, and the orbiting detectors onboard Swift. However, Bromm thinks it's unlikely that the GRB 050904 was a Population III star.

Bromm and Loeb found that if a Population III was to generate a GRB, it would have to be a member of a binary star system. A binary companion would strip the outer layers of the Population III star leaving less material to block the fierce jet when it explodes and making it more likely to be visible across the universe.

Above: Shown in this artist's concept, the first stars in the universe were behemoths that consumed their available fuel at a prodigious rate, dying quickly and explosively. NASA's Swift satellite may detect the resulting gamma-ray bursts, opening a new window onto the early history of the cosmos. Image courtesy David A. Aguilar (Harvard-Smithsonian Center for Astrophysics).

"If Pop III binaries are common, Swift will be the first observatory to probe Population III star formation at high redshifts," said Loeb.

http://www.cfa.harvard.edu/press/pr0531.html

http://arxiv.org/abs/astro-ph/0509303

22 September 2005
Tycho's Remnant Provides Shocking Evidence for Cosmic Rays

The origin of cosmic rays has been a long standing mystery. They are the nuclei of atoms traveling at near the speed of light and strike the Earth's atmosphere from all directions. They generate a cascade of secondary particles in our atmosphere.

Now the Chandra X-ray observatory provides new evidence that the shockwave in the Tycho supernova remnant in Cassiopeia is the origin of many cosmic rays. It supports the long standing idea that supernovae produce most cosmic rays.

This conclusion comes from comparing theoretical models of supernova explosions to the actual observations. Typical models of supernova explosions show that the outward-moving shock wave lies about two light-years ahead of the stellar debris. What Chandra found with the Tycho supernova remnant is that the gap is only one-quarter of this distance. Energy from the outward moving shock must be going elsewhere.

"The most likely explanation for this behavior is that a large fraction of the energy of the outward-moving shock wave is going into the acceleration of atomic nuclei to speeds approaching the speed of light," said Jessica Warren, also of Rutgers University, and the lead author of the report in the Astrophysical Journal.

"With only a single object involved we can't state with confidence that supernova shock waves are the primary source of cosmic rays," said John P. Hughes of Rutgers University. "What we have done is present solid evidence that the shock wave in at least one supernova remnant has accelerated nuclei to cosmic ray energies."

http://chandra.harvard.edu/press/05_releases/press_092205.html

http://xxx.lanl.gov/abs/astro-ph/0507478

27 September 2005
Spitzer and Hubble Find a "Big Baby" Galaxy in the Newborn Universe

One of the farthest and most massive galaxies ever detected has been found using images from the Hubble and Spitzer Space Telescopes. In fact, Hubble's optical cameras couldn't see

the distant galaxy because most of its light comes to us in the form of infrared. NICMOS did see a faint trace of the galaxy, but the Spitzer found it easily, as did an infrared camera on the Very Large Telescope (VLT) at the European Southern Observatory

The galaxy, called HUDF-JD2, was expected to be a young baby galaxy like many others found at such a large distance from Earth. The light from this galaxy left when the Universe was only about 800 million years old. It was found in the Hubble Ultra Deep Field.

"This galaxy appears to have 'bulked up' amazingly quickly, within a few hundred million years after the Big Bang," says Bahram Mobasher of the European Space Agency and the Space Telescope Science Institute. "It made about eight

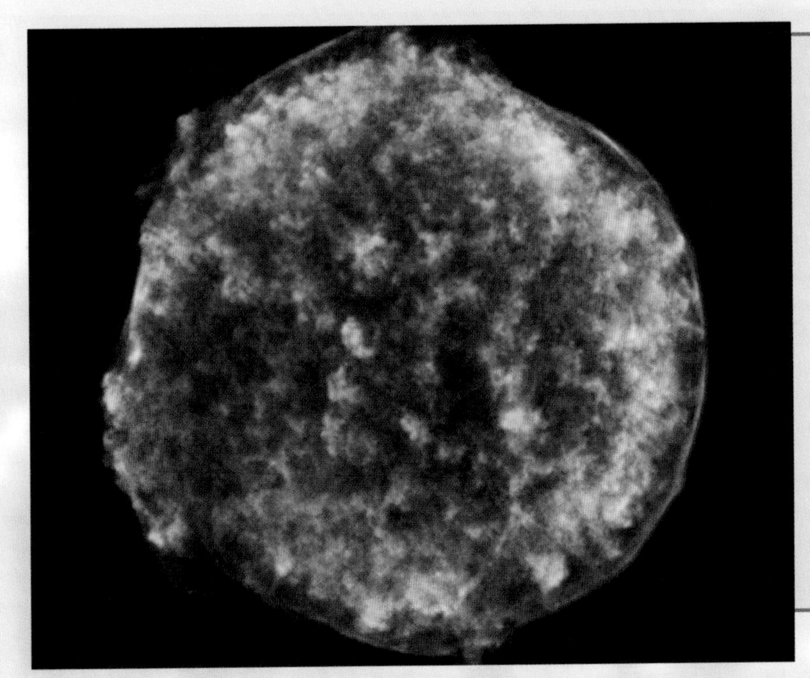

■ **Above:** *Gamma-ray bursts shine so brightly that astronomers can spot them across the universe, as shown in this artist's concept. It is predicted that approximately one-tenth of all bursts captured by Swift will come from stars that died during the first one billion years of the universe. Image courtesy David A. Aguilar (Harvard-Smithsonian Center for Astrophysics).*

■ **Left:** *In 1572, the Danish astronomer Tycho Brahe studied the explosion of a star that became known as Tycho's supernova. More than four centuries later, Chandra's image of the supernova remnant shows an expanding bubble of multimillion degree debris (green and red) inside a more rapidly moving shell of extremely high energy electrons (filamentary blue). Image courtesy NASA/CXC/Rutgers/J.Warren and J.Hughes et al.*

times more mass in terms of stars than are found in our own Milky Way today, and then, just as suddenly, it stopped forming new stars. It appears to have grown old prematurely."

http://hubblesite.org/news/2005/28

http://www.spitzer.caltech.edu/Media/releases/ssc2005-19/

http://xxx.lanl.gov/abs/astro-ph/0509768

29 September 2005
ESO's VLT Captures Image of Spiral Galaxy NGC 1350

The spectacular spiral galaxy, NGC 1350, has been imaged by ESO's Very Large Telescope in a 16-minute exposure. The galaxy, located 85 million light years away in the constellation of Fornax, is classified as a Sa(r) type galaxy, indicating it's a spiral with large central ring.

The galaxy is slightly larger than our Milky Way, spanning 130,000 light years. The outer spiral arms originate from the central ring. Many more distant galaxies can be seen through the tenuous outer arms of the galaxy. The blue color of the arms comes from very young and massive stars. Fine tendrils of dust, in reality massive collections of dark dust similar to those in our Milky Way, create a web-like pattern across the inner ring. This material represents some of the raw material that will form new stars.

http://www.eso.org/outreach/press-rel/pr-2005/phot-31-05.html

Above: *Data from the Spitzer and Hubble Space Telescopes identify one of the most distant galaxies ever seen. This galaxy is unusually massive for its youthful age of 800 million years. [Left] The galaxy, named HUDF-JD2, was pinpointed among approximately 10,000 others in the Hubble Ultra Deep Field (HUDF). [Upper Right] A blow-up of one small area of the HUDF shows where the distant galaxy is located (inside green circle). [Center Right] The galaxy was detected using Hubble's Near Infrared Camera and Multi-Object Spectrometer (NICMOS). It appears very faint and red. [Bottom Right] The Spitzer Infrared Array Camera (IRAC), easily detects the galaxy at longer infrared wavelengths. Images courtesy NASA, ESA, B. Mobasher (STScI and ESA).*

Below: *A colour-composite image of the spiral galaxy NGC 1350 taken with the FORS2 imstrument on the ESO Very Large Telescope. The image, totalling 16 minutes of observations, clearly reveals the delicate structures in this gigantic 'eye' as well as many background galaxies. Image courtesy ESO, with thanks to Henri Boffin, Haennes Heyer and Ed Janssen.*

OCTOBER 2005

4 October 2005

Chandra and Hubble Glimpse Faint Afterglow Of Nearby Stellar Exploration

The Hubble Space Telescope and Chandra X-ray Observatory have combined their data to generate a new view of a distant and faint supernova remnant.

The faint remnant of a 3,000 year old supernova is located 160,000 light years away in the Large Magellanic Cloud. The complex bubble of material, called N132D, is the result of the expanding shock wave from the supernova striking interstellar gas. Hubble's color filters allow details of hydrogen and oxygen emission to be isolated.

Chandra's X-ray image reveals a large horseshoe shaped nebula that has been heated to over 10 million degrees Celsius (18 million degrees Fahrenheit). This high temperature is generated when the supernova shock wave, traveling at 2,000 kilometers per second strikes interstellar gas.

http://hubblesite.org/news/2005/30

http://chandra.harvard.edu/press/05_releases/press_100405.html

5 October 2005

The 30-year Short Gamma-Ray Burst Mystery Finally Solved.

The 35-year-old mystery of the origin of short gamma-ray bursts is finally solved. The powerful, split-second flashes have been too fast for follow-up observations until now. Using the orbiting Swift and HETE-2 satellites, combined with a large coordinated effort of ground-based telescopes, scientists ruled out the option of massive stellar explosions called hypernovae in favor of the merger of two very compact objects. The collisions are either between a black hole and a neutron star or between two neutron stars. The final result of both collisions is a new black hole.

"Gamma-ray bursts in general are notoriously difficult to study, but the shortest ones have been next to impossible to pin down," said Neil Gehrels, principal investigator for the Swift satellite at NASA's Goddard Space Flight Center. "All that has changed. We now have the tools in place to study these events," he said.

Below: *A glowing horseshoe-shaped cloud of hot gas against a backdrop of thousands of stars in the Large Magellanic Cloud. Observations with Chandra (X-ray/blue) and Hubble (optical/pink & purple) were used to make this composite image of N132D, a supernova remnant that was produced by the explosion of a massive star. Images courtesy NASA/SAO/CXC – X-ray, and NASA, ESA, The Hubble Heritage Team (STScI/AURA) – Optical.*

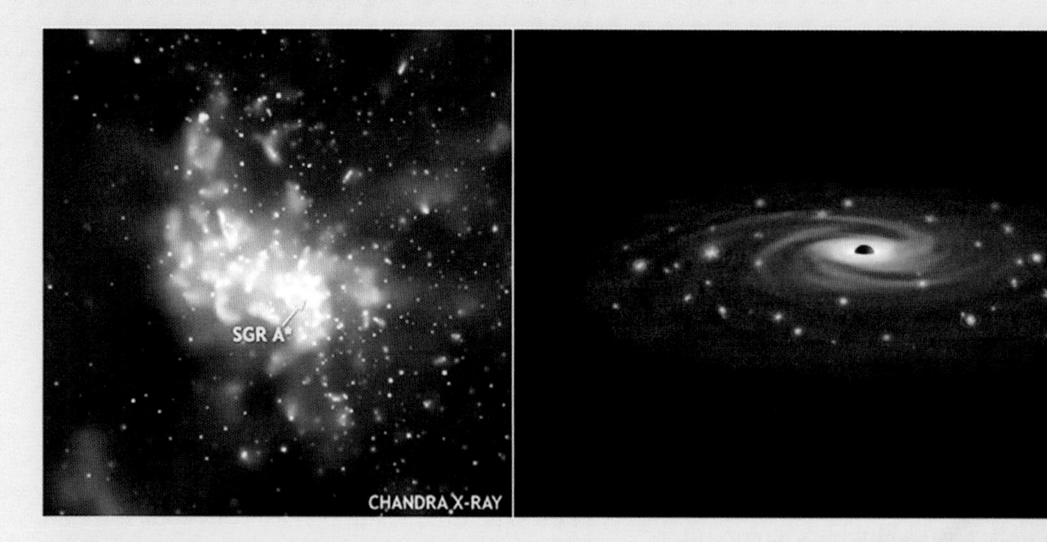

Two short bursts occurred in May and July of 2005. GRB 050509B was seen on May 9 by the Swift satellite and for the first time determined the position of a short burst. GRB 050709, a 70-millisecond burst, was caught by HETE-2. Just over a day later, Jens Hjorth and his team swung the Danish 1.5m telescope at ESO La Silla to find the presence of a fading source sitting on the edge of a galaxy.

"We have thus discovered the first optical afterglow of a short gamma-ray burst", says co-author Kristian Pedersen, of the University of Copenhagen.

The GRB 050509B was found near a galaxy that only contained old stars. Hypernovae are the explosions of massive young stars. This one observation gave credence to the alternative theories for origin of short GRB's.

"We had a hunch that short gamma-ray bursts came from a neutron star crashing into a black hole or another neutron star, but these new detections leave no doubt," said Dr. Derek Fox at Penn State University. Fox led a team that discovered the X-ray afterglow of the July 9 burst with the Chandra X-ray Observatory.

Mergers of black holes and neutron stars are expected to generate create gravitational waves, ripples in space-time predicted by Einstein. New instruments now active, such as the Laser Interferometer Gravitational-Wave Observatory (LIGO), should be able to detect such waves if a short GRB takes place within range.

[Editor: Such dramatic discoveries are rare – in this book you will find major invited feature articles about GRB's, the Swift satellite, and the new LIGO observatory, written by leading researchers in these fields.]

http://www.nasa.gov/mission_pages/swift/bursts/short_burst_oct5.html

http://swift.gsfc.nasa.gov/docs/swift/swiftsc.html

http://xxx.lanl.gov/abs/astro-ph/0510110

http://xxx.lanl.gov/abs/astro-ph/0510098

http://xxx.lanl.gov/abs/astro-ph/0506104

http://xxx.lanl.gov/abs/astro-ph/0510190

http://xxx.lanl.gov/abs/astro-ph/0510096

13 October 2005
Chandra Reveals New Star Generation

A surprising new generation of stars in close proximity to the Milky Way's supermassive black hole, located in Sagittarius, has been revealed by the Chandra X-ray observatory. The results offer a solution to a long-standing puzzle between two competing models of how stars could form so close to a black hole.

Current theories of star formation say that tidal forces around a black hole are so strong that are so strong that few stars can form in such a hostile region. In the migration model, stars formed farther away from the central black hole migrate inwards. This idea predicts about a million low mass stars in the inner region studied by Chandra. In a competing model, the mass of the dust and gas disk counteracts the tidal forces of the black hole, allowing stars to form. However, lower mass stars will be disrupted, giving preference to

***Above:** Chandra's image of the Galactic Center (left) has provided evidence for a new and unexpected way for stars to form. A combination of infrared and X-ray observations indicates that a surplus of massive stars has formed from a large disk of gas around Sagittarius A*, the Milky Way's central black hole (illustration on right). Images courtesy NASA/CXC/MIT/F.K.Baganoff et al. – X-ray, and NASA/CXC/M.Weiss – artist's concept.*

higher mass star formation. Consequently fewer low mass stars will be found.

By comparing X-ray emission from the Orion star forming region with the data around SgrA* (pronounced "Sagittarius A star" – the object at the center of our galaxy suspected of being a supermassive black hole), Chandra found about 10,000 low mass stars, a number that is too low for the migration model and thus ruling it out.

"In one of the most inhospitable places in our galaxy, stars have prevailed," Sergei Nayakshin of the University of Leicester in England said. "It appears star formation is much more tenacious than we previously believed."

"We can say the stars around SgrA* were not deposited there by some passing star cluster, rather they were born there," Rashid Sunyaev of the Max Plank Institute for Physics in Garching, Germany said. "There have been theories that this was possible, but this is the first real evidence. Many scientists are going to be very surprised by these results."

(Editor: See 20 September for news about hot young stars around the central black hole in M31.)

http://chandra.harvard.edu/press/05_releases/press_101305.html

http://xxx.lanl.gov/abs/astro-ph/0507687

http://xxx.lanl.gov/abs/astro-ph/0601268

13 October 2005
Spitzer Sees Turmoil Next Door In "Tranquil" Andromeda Galaxy

The Andromeda Galaxy (M31) turns out to harbor great turmoil in a new view by the Spitzer Space Telescope. In the sharpest infrared images ever taken of the galaxy, details reflect theoretical predictions that a satellite dwarf galaxy passed all the way through M31 a few million years ago.

While visible images show a beautiful and apparently tranquil spiral galaxy, Spitzer is able to trace warm dust almost all the way to the center of the galaxy, where the spiral arms originate. "That's a classic signature of spiral galaxies, but it's not been seen before in M31. And we've traced the spiral arms -- actually they're arm segments -- out beyond the star-forming ring," said Karl D. Gordon of the Steward Observatory in Arizona.

The star-forming ring seen previously is offset from the center of the galaxy and reveals a split in one location. It is this feature that indicates M32, one of M31's satellite galaxies, once traveled through the plane of M31.

"What we're really seeing looks like a hole with a rim, like the first ripple flowing out when you toss a stone in a pond," said George Rieke of The University of Arizona.

Above: *NASA's Spitzer Space Telescope has obtained the sharpest composite image ever taken of dust in the Andromeda galaxy. Asymmetrical features, seen in the prominent ring of star formation, may have been caused by satellite galaxies around Andromeda as they plunge through its disk. Spitzer reveals delicate tracings of spiral arms within this ring that reach into the center of the galaxy. The traditional view at visible wavelengths (lower left), shows the starlight instead of the dust. The multi-wavelength view (lower right) allows astronomers to measure the temperature of the dust by its color. Images courtesy NASA/JPL-Caltech/K. Gordon (University of Arizona) and NOAO/AURA/NSF.*

Extensive computer simulations carried out by researchers at Swinburne University in Australia revealed that the features seen in the Spitzer images could be reproduced by such a collision.

http://www.spitzer.caltech.edu/spitzer/

http://dirty.as.arizona.edu/~kgordon/m31_press/m31_press.html

http://xxx.lanl.gov/abs/astro-ph/0601314

17 October 2005
Feeding the Monster - New VLT Images Reveal the Surroundings of a Supermassive Black Hole

NGC 1097 is a fine barred spiral galaxy lying 45 million light years away in the constellation of Fornax. The core of the galaxy is a particularly active star-forming region. In a new look with adaptive optics on ESO's Very Large Telescope the finest image of the central region of the galaxy has been obtained. The resolution achieved with the images is about 0.15 arcsecond, corresponding to about 30 light-years across.

The image reveals details of a complex central network of filamentary structure spiraling down to the centre of the galaxy, a region where a supermassive black hole lurks.

"This is possibly the first time that a detailed view of the channeling process of matter, from the main part of the galaxy down to the very end in the nucleus, is released," says Almudena Prieto (Max-Planck Institute, Heidelberg, Germany).

Astronomers used a masking technique to reveal the dusty spiral in the central region. The curling of the spiral pattern in the innermost 300 light-years seems to confirm the presence of a super-massive black hole in the centre of NGC 1097. Not seen in the image is the central dusty torus or gaseous disk expected to be seen in active galactic nuclei like NGC 1097 – but these observations place an upper limit of 10 parsecs diameter for such a disk or torus.

http://www.eso.org/outreach/press-rel/pr-2005/phot-33-05.html

http://xxx.lanl.gov/abs/astro-ph/0507071

Below: *The left image shows the central region of the active galaxy NGC 1097 at near-infrared wavelengths. It reveals the nucleus, the spiral arms extending up to 1,300 light-years from the centre, and the star-forming ring comprising more than 300 star-forming regions. The right image shows the same, but after a masking process has been applied to suppress the central stellar light of the galaxy. This unveils a complex central network of filamentary structures spiralling down to the region at the centre where a supermassive black hole probably lurks. Images courtesy ESO and Almudena Prieto (Max-Planck Institute, Heidelberg, Germany) et al.*

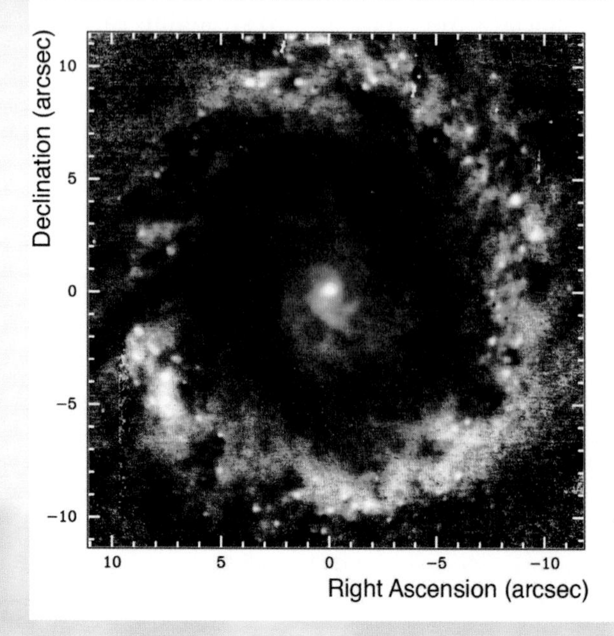

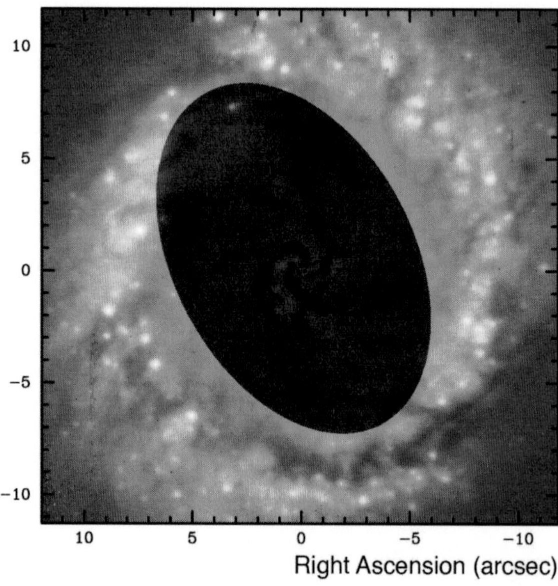

26 October 2005
Large Binocular Telescope Successfully Achieves "First Light"

The Large Binocular Telescope (LBT) achieved "First Light" with one of its mirrors on 12 October 2005. The first released image of the edge-on spiral galaxy, NGC 891, reveals exquisite detail.

■ The "first light" image from the Large Binocular Telescope atop Mount Graham in Arizona. It shows the spiral galaxy NGC891 with a "cool blue" color display. Image courtesy Large Binocular Telescope Observatory.

"First Light" is always a significant milestone for any telescope. The unique instrument houses two enormous 8.4-meter (27.6 foot) primary mirrors mounted side-by-side to produce a telescope equivalent of an 11.8-meter (39 foot) aperture. In addition, the two mirrors can be combined as an interferometer, generating resolution equivalent of a 22.8-meter telescope. The LBT is a collaborative project between numerous astronomy and academic institutions in the U.S.A., Italy and Germany.

(Editor: See the invited feature article by the Director of the LBT, Dr Richard Green, and Technical Director, John Hill, which gives full details of this magnificent new telescope, later in this volume.)

http://medusa.as.arizona.edu/lbto/

NOVEMBER 2005

1 November 2005

New Distance to the Andromeda Galaxy

An eclipsing binary star in the Andromeda galaxy has led to the first determination of the galaxy's distance independent of the Cepheid distance scale. The Cepheid scale has relied on the distance to other galaxies such as the LMC. The use of an eclipsing binary is a more direct method. Because each star passes behind the other over a known period, the precise timing of the occultation leads to the physical dimensions of each star, and hence the luminosity of each star. When compared with each star's apparent magnitude, the distance is calculated using the inverse square law. The Andromeda Galaxy is 2.52 ± 0.14 million light-years from Earth using this technique. This compares well with

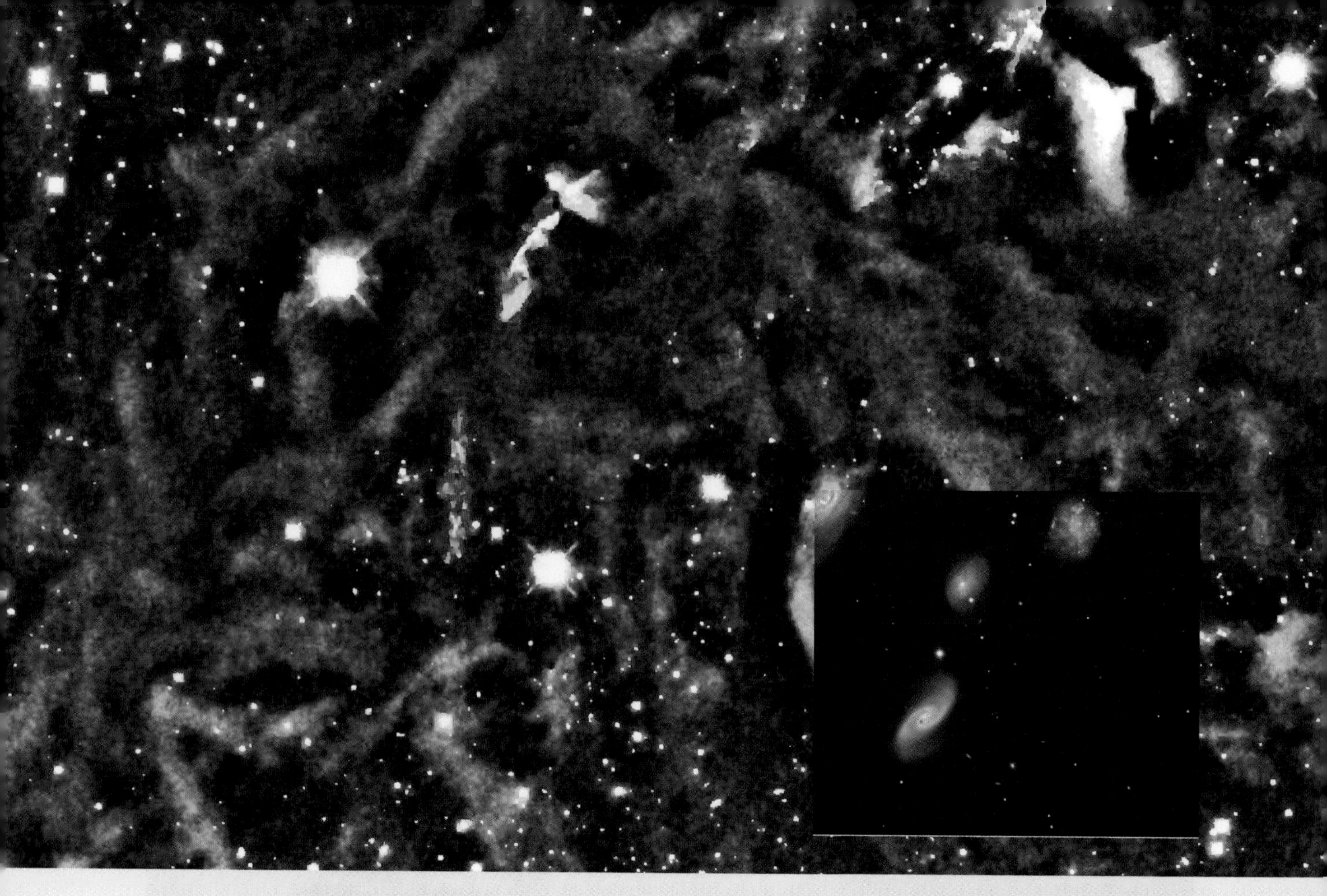

the distance using Cepheids, and increases confidence in the Cepheid distance scale for more distant galaxies.

http://arxiv.org/abs/astro-ph/0511045

3 November 2005
Cosmic Cloudshine: Its Beauty Is More Than Skin Deep

It used to be thought that dark interstellar clouds were really black – they certainly look that way in optical images. Infrared images allow astronomers to peer inside these dark clouds to see the buried stellar nurseries nestled within them. But recent ground-based infrared images of dark clouds have revealed something new, akin to clouds on Earth glowing by reflected streetlights. The otherwise "dark" clouds are illuminated by faint starlight, and long exposure images reveal this ethereal glow. Dubbed "Cloudshine" by Harvard astronomers, the discovery came from part of the COMPLETE survey, (Coordinated Molecular Probe Line Extinction Thermal Emission) of star-forming regions. This survey is imaging three stellar nurseries at high resolution and over wide angles.

"Other astronomers have seen hints of "cloudshine" in their images, but our new photographs are the most spectacular evidence of "cloudshine" to date," said Alyssa Goodman of the Harvard-Smithsonian Center for Astrophysics (CfA).

http://www.cfa.harvard.edu/press/pr0534.html

http://arxiv.org/abs/astro-ph/0510624

http://arxiv.org/abs/astro-ph/0510600

__Above:__ The green billows in this color-coded near-infrared image of the star-forming region L1448 show newly discovered 'cloudshine', which offers a way of peering below the surface of dark nebulae to map star-forming regions in exquisite detail. The red streaks are jets of material ejected from protostars during the accretion process. This image was taken with the OMEGA 2000 camera at the Calar Alto Observatory in Spain as part of the COMPLETE survey of star-forming regions. Image courtesy J. Foster and A. Goodman (Harvard-Smithsonian Center for Astrophysics).

4 November 2005
Cosmic Portrait of a Perturbed Family

One of the finest examples of compact groups of galaxies has been captured by the European Southern Observatory's Very Large Telescope on Cerro Paranal, Chile. Robert's Quartet, located about 160 million light-years away, contains four very different galaxies. They lie within the constellation of the Phoenix. Such groups are excellent regions to study the interplay of gravity as these four galaxies interact with each other.

Its members are NGC 87, NGC 88, NGC 89 and NGC 92, and they lie within 75,000 light years of each other. At the distance of the group, this spans an angle of 1.6 arcminutes.

The VLT has studied star-forming (HII) regions in each galaxy, and found the most active is NGC 92, with over 200 HII regions, some along the 100,000 light year long disrupted spiral arm, while NGC 87 revealed just over 50. The other two galaxies, NGC 88 and 89, had very few HII regions. Yet NGC 88 reveals an interesting plume feature, and NGC 89 shows a ring of new stars, so the gravitational interactions between these two are having their effects.

http://www.eso.org/outreach/press-rel/pr-2005/phot-34-05.html

Left: *An image of the group of galaxies known as Robert's Quartet from data collected by the FORS2 instrument on ESO's Very Large Telescope. The quartet is one of the finest examples of compact groups of galaxies. Because such groups contain several galaxies in a very small region, they are excellent laboratories for the study of galaxy interactions and their effects, in particular on the formation of stars. Image courtesy ESO, with thanks to Henri Boffin, Kristina Boneva and Haennes Heyer.*

Below: *Here, an infrared image taken by NASA's Spitzer Space Telescope is compared with a visible-light picture of the same region (inset). While the infrared view reveals towering pillars of dust aglow with the light of embryonic stars (white/yellow), the visible-light view shows dark, barely-visible pillars. Spitzer is both seeing, and seeing through, the dust. Images courtesy NASA/JPL-Caltech/L. Allen (Harvard-Smithsonian Center for Astrophysics) and California Institute of Technology's Digitized Sky Survey – Visible.*

9 November 2005
Spitzer Captures Cosmic 'Mountains Of Creation'

A new infrared image from the Spitzer Space Telescope reveals the inner workings of a huge interstellar cloud of gas and dust. The image, as iconic as the famous Hubble image of the Eagle Nebula called the "Pillars of Creation", shows huge "mountains" of glowing warm dust that are ten times the size of the pillars in the Eagle Nebula.

The object is called W5 and lies 7,000 light years away in the constellation of Perseus. Dozens of new stars forming within the cloud, previously veiled by opaque dust, are revealed by the infrared view. Radiation and massive stellar winds have sculpted the cloud of material from inside. The fingers of dust appear to radiate away from a point outside the image – that location contains a massive star thought to have triggered additional star formation in this massive cloud.

The image represents more than just a pretty picture. "With Spitzer, we can not only see the stars in the pillars, but we can estimate their age and study how they formed," said Joseph Hora, from the Harvard-Smithsonian Center for Astrophysics.

Above: *This image of the young star cluster NGC 346 and its surrounding star formation region, located in the Small Magellanic Cloud, was taken with the Hubble Space Telescope's Advanced Camera for Surveys. The NGC 346 cluster is resolved into at least three sub-clusters and contains dozens of hot, blue, high-mass stars. A torrent of radiation from these hot, young stars eats into denser areas creating a fantasy sculpture of dust and gas. Image courtesy NASA, ESA, and Antonella Nota (STScI/ESA)*

http://www.spitzer.caltech.edu/Media/releases/ssc2005-23/release.shtml

http://xxx.lanl.gov/abs/astro-ph/0507705

10 November 2005
Hubble Views Powerful Outflows Sculpting Gas Around Young Stars

Details of a bright star cluster in our neighboring galaxy, the Large Magellanic Cloud (LMC), have been revealed in a new image by the Hubble Space Telescope. NGC 346 lies 210,000 light years away, and Hubble has acquired the clearest view ever of the region. Within the image, a new cluster of infant stars has been discovered.

Radiation and stellar winds from the hot stars in the cluster have sculpted weird and fascinating structures within the surrounding gas and dust. In places, dark nodules appear as tear-drop shapes as they are streamlined by passing stellar winds.

A team of astronomers led by Dr. Antonella Nota of the European Space Agency and the Space Telescope Science Institute has used the data to discover a rich population of infant stars within NGC 346. Their ages are between 3 to 5 million years old, but have not yet begun hydrogen burning. The locations of the infant stars are coincident with observed molecular clouds detected by the Infrared Space Observatory (ISO), and provide valuable clues to understanding how star formation has progressed in this region of the LMC.

http://hubblesite.org/newscenter/newsdesk/archive/releases/2005/35/

http://xxx.lanl.gov/abs/astro-ph/0602218

17 November 2005
Einstein's Rings in Space

The Sloan Digital Sky Survey, in combination with the Hubble Space Telescope, has produced a 20% increase in the number of known Einstein rings. The total now stands at 119. They represent the latest results from the Sloan Lens ACS Survey (SLACS). (ACS is the camera on board Hubble).

Einstein Rings are produced when the light from a very distant galaxy passes close to an intervening galaxy whose gravity distorts the image of the more distant cousin. Acting like a lens, the intervening galaxy creates multiple images of the distant galaxy, often in the form of arcs, and in the case of perfect alignment, complete rings. Out of the 19 new objects, 8 are

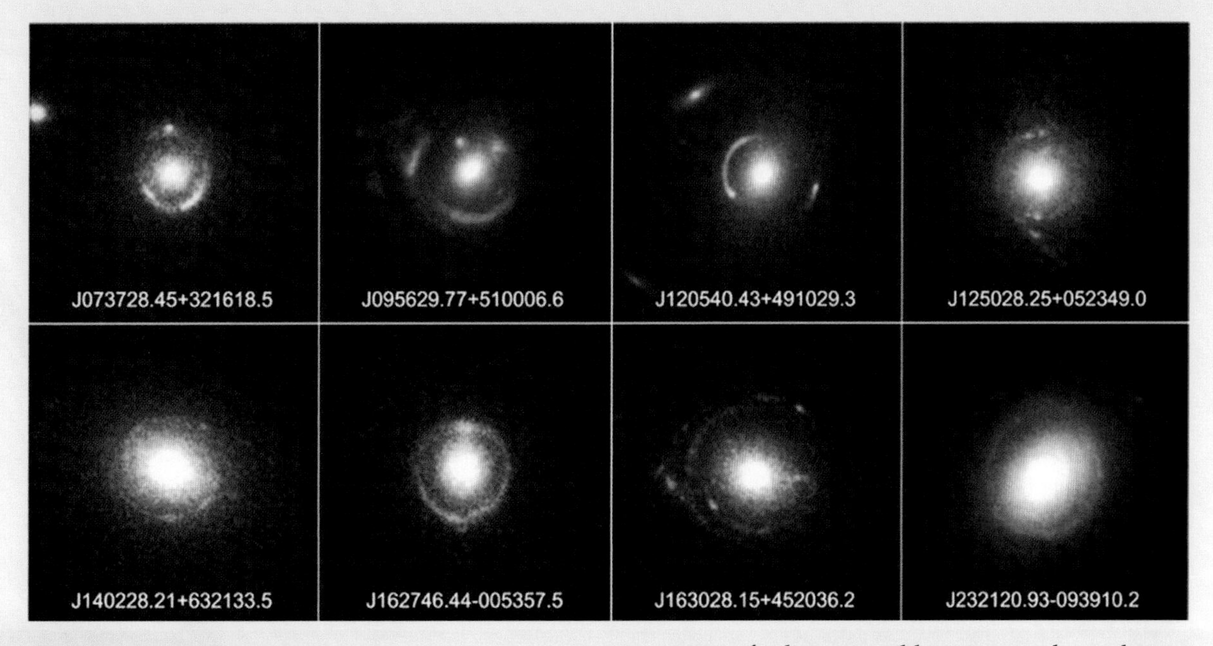

complete rings, more than tripling the previous number known.

Predicted by Einstein in an application of his General Theory of Relativity, the amount of distortion in the Einstein Ring reveals amount of mass in the intervening galaxy. It has been known for decades that the amount of material in a galaxy far outweighs the visible portion. The Einstein rings give a valuable way to investigate the distribution of dark matter in more galaxies, and offer further insight into the process of galaxy formation.

"An Einstein ring is one of the most dramatic demonstrations of the General Theory of Relativity in the cosmos," said Adam Bolton of the Harvard-Smithsonian Center for Astrophysics (CfA). "It provides a unique opportunity to study the most massive galaxies in the universe."

http://www.cfa.harvard.edu/press/pr0537.html

http://hubblesite.org/news/2005/32

http://www.slacs.org

http://www.arxiv.org/abs/astro-ph/0511453

30 November 2005

HARPS Instrument Finds Neptune-Mass Exoplanet Around Small Star

The discovery of a Neptune-mass planet orbiting a small red dwarf star raises the possibility of many more planets around the most common stars in our galaxy.

Discovered using the HARPS instrument on the 3.6-m telescope at La Silla (Chile), the planet was revealed by the wobble it induces on its parent star. The data revealed a planet 17 times the mass of the Earth orbiting the red dwarf star, Gleise 581, every 5.3 days. This places the planet a mere 6 million kilometers from the tiny star.

"Our finding possibly means that planets are rather frequent around the smallest stars," says Xavier Delfosse, from the Laboratoire d'Astrophysique de Grenoble (France). "It certainly tells us that red dwarfs are ideal targets for the search for exoplanets."

Gleise 581 lies 20.5 light years from Earth in the constellation of Libra. There are 80 similar stars among the closest 100 to our Sun, although a recent survey of about 200 red dwarfs turned up only two planets.

http://www.eso.org/outreach/press-rel/pr-2005/pr-30-05.html

http://xxx.lanl.gov/abs/astro-ph/0509211

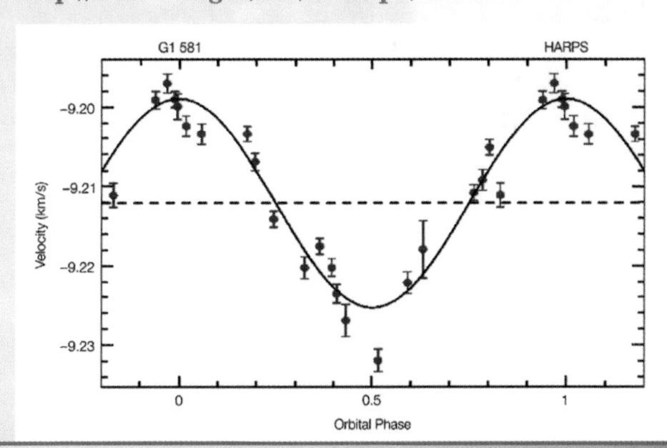

Top: The thin blue bull's-eye patterns in these six Hubble Space Telescope images are so-called 'Einstein rings' – an elegant manifestation of gravitational lensing. The yellowish-red blobs are giant elliptical galaxies roughly 2 to 4 billion light-years away. Einstein rings are created as the light from galaxies at least twice as far away is distorted into circular shapes by the gravity of the intervening giant elliptical galaxies. Images courtesy NASA, ESA, A. Bolton (Harvard-Smithsonian Center for Astrophysics), and the SLACS Team.

Above: This radial velocity curve for the red dwarf star Gliese 581 shows the wobble induced on the parent star by a newly found planet about 17 times the mass of the Earth. Discovered using the HARPS instrument on the 3.6-meter telescope at La Silla (Chile), it is one of the smallest extrasolar planets found to date. It orbits in only 5.366 days at a mean distance from its star of just 6 million kilometers. Graphic courtesy European Southern Observatory and Stéphane Udry (Geneva Observatory).

DECEMBER 2005

1 December 2005

Most Detailed Image of the Crab Nebula

The Crab Nebula is the finest supernova remnant in the entire sky. After erupting into plain sight in the year 1054, the expanding shell of debris has been glowing ever since, and every major telescope on the planet has, at one time or another, focused on it, and the Hubble Space Telescope is no exception.

Hubble's fine Wide Field and Planetary Camera 2 (WPFC2) has produced the finest resolution image to date of the Crab Nebula. Using a combination of filters that isolate certain wavelengths of light from elements within the nebula, and 24 individual exposures, the image reveals intricate structures within the 6-light-year-wide nebula.

Near the center of the nebula is the pulsar that powers the blue diffuse glow from the nebula. The filaments are a product of the outwardly expanding shell from the supernova having struck previously emitted slow moving gas from the progenitor star long before it exploded.

This image of the Crab Nebula supernova remnant was assembled from 24 exposures taken with the Hubble Space Telescope's Wide Field and Planetary Camera 2. The colors show the different elements expelled during the explosion. Blue indicates neutral oxygen, green singly-ionized sulphur and red doubly-ionized oxygen. The data have been superimposed onto images from the ESO Very Large Telescope. Image courtesy NASA, ESA and Allison Loll/Jeff Hester (Arizona State University), with thanks to Davide De Martin.

Compare this image with the images of SN1987A in the feature article in this volume for an idea of what the nebula probably looked like ten years after it exploded.

Chemical elements heavier than iron are formed in supernovae, and are the most important mechanisms for distributing heavy elements throughout the universe. Such elements eventually become part of planets that form around other stars.

http://www.spacetelescope.org/news/html/heic0515.html

http://arxiv.org/abs/astro-ph/0408061

http://arxiv.org/abs/astro-ph/0112250

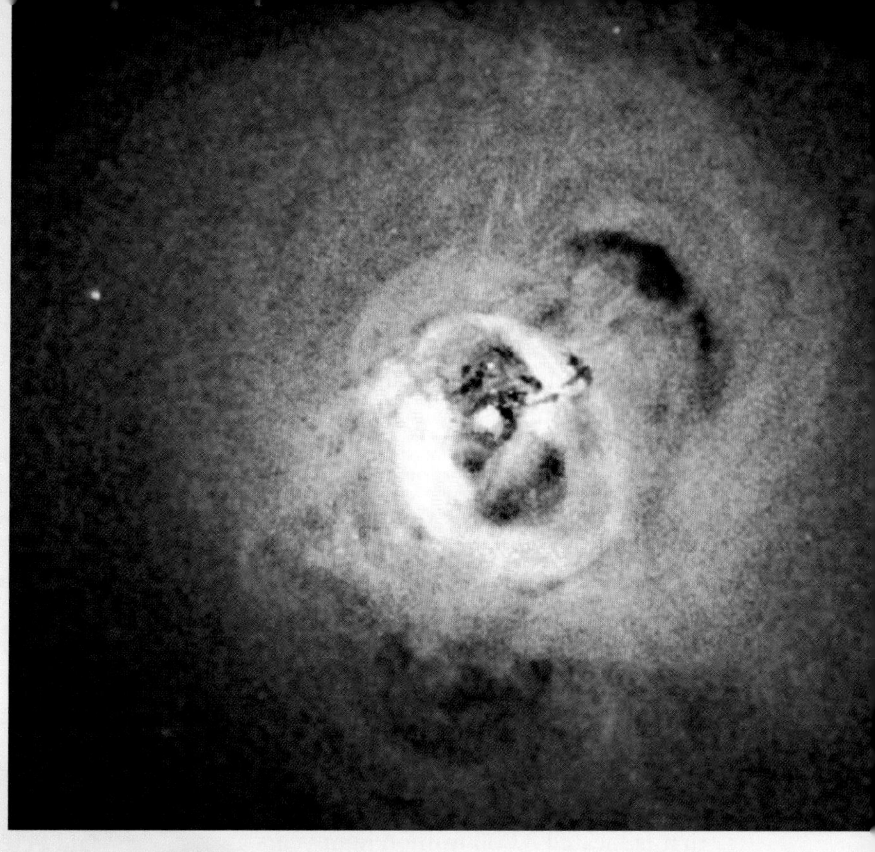

1 December 2005
Chandra Proves Black Hole Influence is Far Reaching

In the longest X-ray observation ever taken of a galaxy cluster, a team of British astronomers using NASA's Chandra X-ray Observatory found massive plumes of material stretching 300,000 light years from their source, a supermassive black hole.

The new findings show the incredible power that supermassive black holes have over intergalactic distances. The observations were made of the Perseus Cluster, which contains thousands of galaxies. The entire region is bathed in a multimillion degree hot gas.

The bubbles created by high energy particles in high speed jets are centered on the massive elliptical galaxy at the center of the Perseus Cluster.

"The plumes show that the black hole has been venting for at least 100 million years, and probably much longer," said co-author Jeremy Sanders of Cambridge University.

http://chandra.harvard.edu/press/05_releases/press_120105.html

http://xxx.lanl.gov/abs/astro-ph/0510476

7 December 2005
ESO Signs Large Contract for Project ALMA

European Southern Observatory has signed the contract that sets the stage for the construction of the Atacama Large Millimeter Array (ALMA). The 25 antennas, each 12 meters diameter, will be located in Chile on the 5,000 meter high Llano de Chajnantor plain in the Atacama Desert. When completed, the array will establish baselines of up to 18 kilometers. The antennas will be built by a European consortium of engineering companies led by Alcatel Alenia Space.

ALMA will achieve sub-arcsecond resolution using the aperture synthesis techniques of radio astronomy, enabling detailed study of dusty regions of space, such was star formation clouds, protoplanetary disks, and the dusty-embedded core of our own galaxy, promising significant advances for astronomy. The array is expected to begin partial operation by 2010-2011.

http://www.eso.org/outreach/press-rel/pr-2005/pr-31-05.html

http://www.eso.org/projects/alma/

Above: *A total of 270 hours of Chandra observations of the central regions of the Perseus galaxy cluster reveals evidence of the turmoil that has wracked the cluster for hundreds of millions of years. The cluster contains thousands of galaxies immersed in a vast cloud of multimillion degree gas. The dark blue filaments in the center are probably a galaxy that has been torn apart and is falling into NGC 1275 (Perseus A), the giant galaxy that lies at the center of the cluster. Image courtesy NASA/CXC/IoA /A.Fabian et al.*

14 December 2005

Black Hole May Have Swallowed a Neutron Star

Short Gamma-Ray Bursts (GRB's) are so powerful, that theorists have decided one of two explanations can account for them. They are thought to arise either from the merger of two neutron stars or of one neutron star with a black hole. A rare event spotted by the Swift GRB observatory on 24 July 2005, appears to have been such an event.

GRB 050724 was one of the best observed short GRB's to date. It lasted only a few milliseconds but the afterglow gave clues to the fate of the neutron star. The afterglow as observed by the orbiting Chandra X-ray Observatory and by the 10-meter Keck telescope on Hawaii. These observations placed the burst at 13,000 light-years away from the centre of an elliptical galaxy that is located 3,000 million light-years away.

"For billions of years this black hole and neutron star orbited each other in a gravitational tug-of-war," said Dr. Scott Barthelmy of NASA's Goddard Space Flight Center in Greenbelt, Md. "The neutron star lost."

The determination of the type of merger comes from the afterglow. It's predicted by current models that a neutron star merger would immediately form a black hole with very little falling into it later, producing a short-lived afterglow. Likewise, two black holes merging would produce little or no afterglow. The long afterglow observed in GRB 050724 indicates that the neutron star broke apart as it merged with the black hole, with pieces of it falling in over a period of time.

"There's only one thing I know of that could rip apart a neutron star with bits flying out, and that's a black hole. Now we have the first evidence that this might actually be occurring," said Peter Meszaros of Penn State University.

http://www.nasa.gov/vision/universe/starsgalaxies/blackhole_meal.html

http://www.eso.org/outreach/press-rel/pr-2005/pr-32-05.html

http://xxx.lanl.gov/abs/astro-ph/0508115

http://xxx.lanl.gov/abs/astro-ph/0511588

http://xxx.lanl.gov/abs/astro-ph/0601661

Below: *This spectrum from NASA's Spitzer Space Telescope shows that some of the most basic ingredients of DNA and protein are concentrated in a dusty planet-forming disk around a young sun-like star called IRS 46. These data also indicate that the ingredients – acetylene and hydrogen cyanide – are located in the star's terrestrial planet zone, the region where scientists believe Earth-like planets would be most likely to form. Image courtesy NASA/JPL-Caltech/F. Lahuis (Leiden Observatory).*

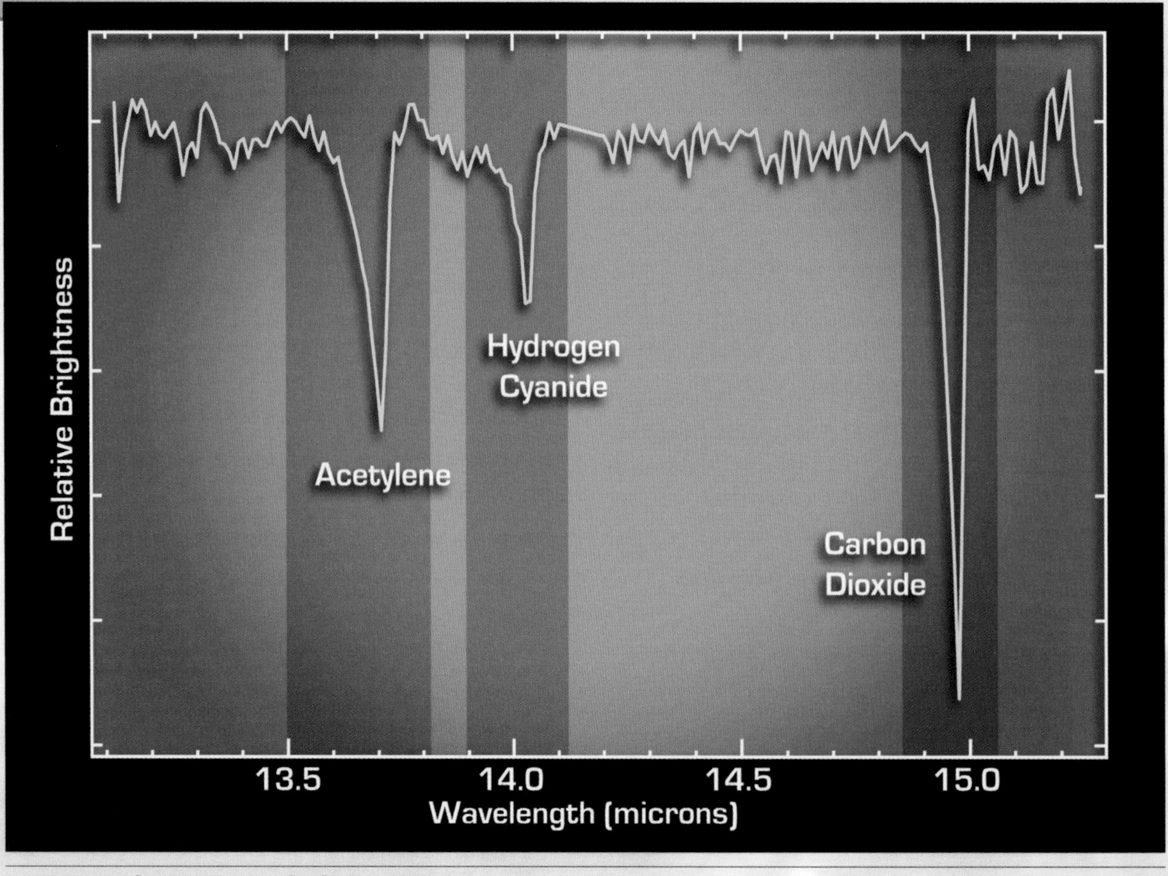

20 December 2005

Precursor to Proteins and DNA Found in Stellar Disk

An infrared source within our own Milky Way Galaxy has revealed signatures of some of the basic compounds necessary to build organic molecules and proteins found within DNA.

The discovery comes from the orbiting Spitzer Space Telescope that observes in the infrared. The molecules were found within the dusty disk surrounding IRS 46, a young stellar object (YSO) embedded in a dusty disk lying in Ophiuchus about 375 light years from Earth. Follow-up observations came from the James Clerk Maxwell Telescope (JCMT) and the Keck telescope, both located on Hawaii.

"We see prebiotic organic molecules in comets and the gas giant planets in our own solar system and wonder, where did these chemicals come from?" said Dr. Marc Kassis, support astronomer at the W. M. Keck Observatory.

Spitzer detected the signatures of acetylene and hydrogen cyanide through the spectrum of IRS46. The same molecules are found within our solar system, such as the atmospheres of the gas planets, and in comets, so it's no surprise that these molecules exits throughout the galaxy, but having direct confirmation of the chemical structure of an object so far away is vital to improving models of planetary formation.

The results come as part of a survey by Spitzer called the "c2d legacy program". It has looked at more than 100 sources in five nearby star-forming regions. IRS 46 is the only one showing clear evidence of any organic compounds in the warm regions inner disk region.

"This infant system might look a lot like ours did billions of years ago, before life arose on Earth," said the science team leader, Fred Lahuis of Leiden Observatory in the Netherlands.

http://www.spitzer.caltech.edu/Media/releases/ssc2005-26/index.shtml

http://arxiv.org/abs/astro-ph/0511786

21 December 2005

Flashes From the Past: Echoes from Ancient Supernovae

When a supernova explodes the light from the brilliant flash travels out at the speed of light, far ahead of any material. Any interstellar dust or gas in the regions around supernovae reflects this light, sometimes back towards the Earth, resulting in so-called "light echoes.

Astronomers using the Mosaic digital camera on the Blanco 4-meter telescope at Cerro Tololo Inter-American Observatory (CTIO) in Chile, have found three such light echoes from old supernovae, all in the same region of the Large Magellanic Cloud (LMC). The supernovae appear to have erupted between 200 and 600 years ago.

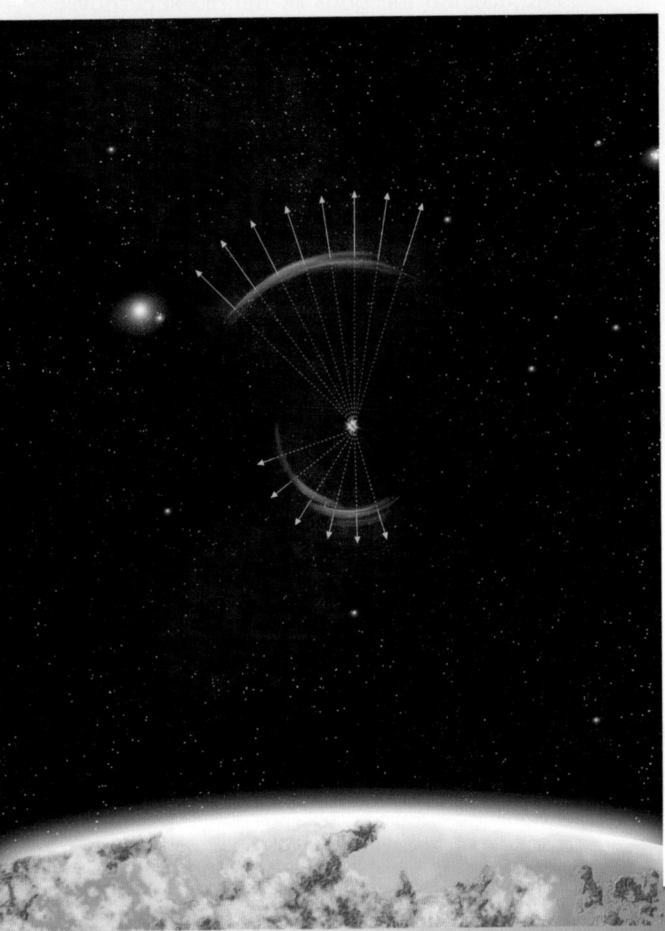

Above: *This artist's concept portrays the appearance of a 'light echo' from a supernova that exploded in the Large Magellanic Cloud, as seen from Earth more than two centuries after the original explosion. The echo is only part of a ring, because to be seen it must intersect with existing clouds of interstellar dust far from the explosion, which are not spaced equally within the large volume into which the supernova's light continues to expand. Illustration courtesy P. Marenfeld and NOAO/AURA/NSF.*

The light echoes were detected by comparing images of the LMC taken years apart. By subtracting one image from another, the concentric arcs of light reflecting off interstellar dust clouds are some of the only features remaining. The technique is being used to detect distortions in light caused by the presence of dark matter in the LMC, part of the SuperMACHO project.

"We see the reflection as an arc because we are inside an imaginary ellipse, with the Earth at one focus of the ellipse and the ancient supernova at the other," explained Nicholas Suntzeff of NOAO. "As we look out toward the supernovae, we see the reflection of the light echo only when it intersects the outer surface of the ellipse. The shape of the reflection from our vantage point appears to be a portion of a circle."

The original locations of the old supernovae are easily determined by figuring out the central focal point of each arc. Previously thought to be young supernovae, the objects have now been found to be hundreds of years older than expected. Team members then fit perpendicular vectors to the curves of each arc system, which were found to point backwards toward the sites of three supernovae remnants, which were previously known and thought to be relatively young.

Seeing the actual light from such old supernovae explosions gives astronomers a unique tool to study the original explosion. "We have the potential with these echoes to determine the star's cause of death, just like the archaeologists who took a CT scan of King Tut's mummy to find out how he died," said co-author Arti Garg of the Center for Astrophysics at Harvard University.

It was back in the 1940's when Fritz Zwicky made the prediction that light from ancient supernova explosions could be observed as echoes from intervening gas clouds. These observations are confirmation of Zwicky's bright idea.

http://www.noao.edu/outreach/press/pr05/pr0512.html

http://arxiv.org/abs/astro-ph/0510738

22 December 2005
Christmas Tree Cluster viewed by Spitzer.

The star-forming region called the Christmas Tree cluster takes on a whole new appearance in an infrared image taken by the Spitzer Space Telescope. The cluster is located 3,000 light years from Earth in the constellation of Monoceros.

The triangular appearance of the cluster as seen through telescopes, with a bright star at its apex, has been christened the Christmas Tree by amateur astronomers.

"Hundreds of new stars and planetary systems have been produced over the past few million years in a prodigious burst of birthing activity within this enormous star-making factory," said Charles

***Right:** Newborn stars, hidden behind thick dust, are revealed in this image from NASA's Spitzer Space Telescope, created using Spitzer's Infrared Array Camera (IRAC) and Multiband Imaging Photometer (MIPS) instruments. IRAC's near and mid-infrared eyes (top right) show that the nebula is still actively forming stars. MIPS' far-infrared eyes (bottom right) reveal the colder dust of the nebula and unwrap the youngest stars from their dusty covering. The newly revealed infant stars appear as pink and red specks toward the center of the combined IRAC-MIPS image (left panel). The stars appear to have formed in regularly spaced intervals along linear structures in a configuration that resembles the spokes of a wheel or the pattern of a snowflake. Hence, astronomers have nicknamed this the 'Snowflake Cluster'. Images courtesy NASA/JPL-Caltech/P.S. Teixeira (Center for Astrophysics).*

Lada of the Harvard-Smithsonian Center for Astrophysics (CfA).

The infrared view reveals previously unseen ribbons of gas and dust swirling among the large number of brilliant stars. The cluster is relatively young and we are viewing inside the dark nebula, seeing for the first time an outline of a giant molecular cloud from which the cluster formed only recently. A new cluster of protostars evenly spaced out within the nebula has been called the Snowflake Cluster. Theoretical models of stellar formation within massive clouds like this predict the spacing between stars is set by the density and temperature of the cloud at the onset of star formation. Such distribution of new protostars provides a unique laboratory to study these effects.

http://www.spitzer.caltech.edu/Media/mediaimages/sig/sig05-028.shtml

http://www.cfa.harvard.edu/press/pr0540image.html

http://arxiv.org/abs/astro-ph/0601300

http://arxiv.org/abs/astro-ph/0511732

27 December 2005

Galaxy's Neighboring Spiral Arm is Closer Than Thought

In past years the measured distance to the Perseus spiral arm has varied between 7,200 -14,000 light years, depending on which technique was applied. Using the Very Long Baseline Array (VLBA), astronomers have now determined the distance at 6,400 light years (with an accuracy of 2%), much closer than previous measurements.

"Our neighbors are closer than we thought," stated first author Ye Xu (Shanghai Astronomical Observatory).

Distance measurements are done using parallax, the variation in a nearby object's position relative to the background caused by the orbital motion of the Earth around the Sun. Very high precision in angular measurements are required. The new VLBA observations achieved 10 microarcsecond accuracy, equivalent to knowing whether a flashlight sits in the left or right hand of an astronaut standing on the Moon as seen from Earth.

The team observed methanol masers inside the region near a newly formed star in the Perseus arm called W3OH. Knowing its accurate distance, astronomers also determined that it is orbiting the galaxy slower than the galaxy spins, and is consequently "falling" inwards toward the center of the galaxy.

"We know less about the structure of our own galaxy than we do about many nearby galaxies like Andromeda," said Mark Reid of the Harvard-Smithsonian Center for Astrophysics. "I have spent more than a decade developing the calibration techniques we needed to obtain this result." Such tenacity has paid dividends.

http://www.vlba.nrao.edu/

http://arxiv.org/abs/astro-ph/0512223

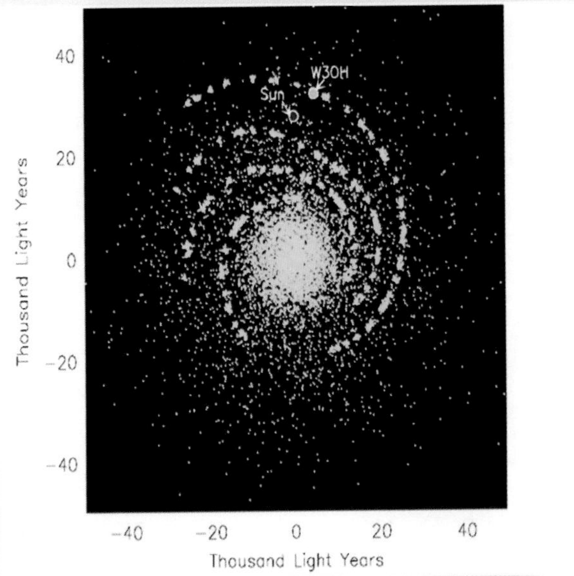

JANUARY 2006

5 January 2006

A Dying Star Reveals More Evidence for a New Kind of Black Hole

There are stellar mass black holes, a few times the mass of the Sun, and supermassive black holes that contain over a million solar masses. But black holes of intermediate mass have been strangely absent. That is, until now.

■ **Left:** *'Plan view' of the Milky Way as seen from its north pole. Estimated locations of spiral arms are indicated by the large numbers of dots and the prominent dashed lines. The locations of the Sun and of the massive star-forming region W3OH in the Perseus spiral arm, are indicated. Graphic courtesy Mark Reid (Harvard-Smithsonian Center for Astrophysics).*

"In the past decade, several satellites have found evidence of a new class of black holes, which could be between 100 and 10,000 solar masses," said Jean Swank, Rossi X-ray Timing Explorer (RXTE) project scientist at NASA's Goddard Space Flight Center, Greenbelt, Md. "There has been debate about the masses and how these black holes would form. Rossi has provided major new insight."

Astronomers using RXTE have now found evidence of a medium sized black hole. The object, called M82 X-1, is part of a group of objects called ultra-luminous X-ray objects. It lies close to a star cluster with over a million stars in a region a mere 100 light years across. Such high density of stars could lead to multiple star collisions where an object might form a black hole in the process and grow to an unusual size. A 62-day period in the X-ray luminosity is probably caused by a companion star orbiting the black hole. The region is dusty, making observations difficult, but it does appear the companion is a large red giant star, large enough to feed matter into the black hole.

"With this discovery of the orbital period, we now have a consistent picture of the whole evolution of a mid-mass black hole binary," said Philip Kaaret of the University of Iowa. "It was formed in a 'super' star cluster; the black hole then captured a companion star; the companion star evolved to the giant stage; and we now see it as an extremely luminous X-ray source because the companion star has expanded and is feeding the black hole."

Other groups have been on the trail of M82 X-1, from Carnegie Mellon University (CMU) in Pennsylvania and from Italy. The CMU group uses data from XMM Newton to determine the mass of this black hole as between 25-520 solar masses. This is in broad agreement with the Italian group who suggest "a few tens to 1,000 solar masses," conforming the identity as an intermediate mass type object.

http://universe.nasa.gov/press/event_horizon/event_horizon.html

http://arxiv.org/abs/astro-ph/0509646

http://arxiv.org/abs/astro-ph/0509796

http://arxiv.org/abs/astro-ph/0409416

5 January 2006
Looking Down the Mouth of an Interstellar Cavern

A cluster of massive stars has excavated a giant mouth-shaped cavity inside a huge cloud of gas and dust, providing astronomers with a wealth

The N44 superbubble complex in the Large Magellanic Cloud, a cloud of gas and dust dominated by a vast bubble about 325 by 250 light-years across, imaged by the Gemini South Telescope in Chile. A cluster of massive stars inside the cavern has cleared away gas to form a distinctive mouth-shaped hollow shell. It is likely that the explosive death of one or more of the cluster's most massive and short-lived stars played a key role in the formation of the large bubble. Composite color image by Travis Rector, University of Alaska Anchorage.

of data and a very dramatic setting. The N44 superbubble, as it is known, may have been formed by one of the massive stars exploding in a supernova. The bubble spans 325 by 250 light years.

"This region is like a giant laboratory providing us with a glimpse into many unique phenomena," said Sally Oey of the University of Michigan, who has studied this object extensively. "Observations from space have even revealed x-ray-emitting gas escaping from this superbubble, and while this is expected, this is the only object of its kind where we have actually seen it happening."

Measured flows of the gas in and around this bubble provide a confusing picture given the expected velocities of winds from the massive stars in the region. The role of supernovae may only be part of the story. Multiple smaller bubbles cling to the edge of the large bubble, suggesting the same process that formed the superbubble is continuing at a smaller scale.

N44 is located about 150,000 light years away in the Large Magellanic Cloud, a satellite galaxy to the Milky Way.

http://www.gemini.edu/

Editor's Note: The following four days of news were released at the 207th meeting of the American Astronomical Society in Washington D.C. from 9-12 January 2006.

9 January 2006
Huge Images Show Majestic Beauty and Violence of Large and Small Magellanic Clouds

The Magellanic Cloud Emission Line Survey (MCELS) has produced more than 2000 images to create large mosaics of the Large and Small Magellanic Clouds (LMC and SMC). The five-year survey used 0.9-meter Curtis Schmidt telescope at the Cerro Tololo Inter-American Observatory in Chile. Five specialized filters isolated light from hydrogen, oxygen, and sulfur, in addition to red and green broad band filters.

"MCELS reveals interstellar gas that has been heated and energized by stars, and so it's an especially valuable data set for understanding how energy and matter cycle between stars and gas," said MCELS team member Sally Oey of the University of Michigan, Ann Arbor.

The LMC image contains 1,500 individual images, with the remaining 500 making up the SMC image.

"In conjunction with surveys of the Magellanic Clouds carried out at other wavelengths, MCELS will lead to a fuller understanding of processes that shape the interstellar medium, and mark the beginnings and ends of stellar lives," said team member Sean Points of NOAO.

http://www.noao.edu/outreach/press/pr06/pr0601.html

10 January 2006
Monster Black Holes Grow After Galactic Mergers

The Hubble Ultra Deep Field (HUDF) provides a unique window on the evolution of galaxy mergers and how the growth of supermassive black holes develops. The results suggest that monster black holes weren't born massive but grew from mergers.

Above: *The glowing gas of the interstellar medium (ISM) is the breeding ground for the formation of new stars, and the cemetery where the ashes of dead stars ultimately return. A new study called the Magellanic Cloud Emission Line Survey (MCELS) has focused on the ISM in the Small and Large Magellanic Clouds. Image courtesy F. Winkler/Middlebury College, the MCELS Team, and NOAO/AURA/NSF.*

"By studying distant galaxies in the Hubble Ultra Deep Field (HUDF), we have the first statistical evidence that supermassive black-hole growth is linked to the process of galaxy assembly," said astronomer Rogier Windhorst, of Arizona State University in Tempe, Arizona. "The HUDF provides an actual look back in time to see snapshots of early galaxies so that we can study them when they were young."

Astronomers searched the HUDF for "tadpole galaxies," so-called because they have bright knots and tails caused by mergers, and found 165, representing about 6% of the 2,700 galaxies in the study.

"To our surprise, however, these tadpole objects did not show any fluctuation in brightness," said Amber Straughn of Arizona State University who led the study. This result indicates that any clumps of material falling into the black hole resulting in variations in light are shrouded from view by dust.

Two distinct phases in galaxy evolution are seen in the HUDF. The early phase produces the so-called tadpole stage. In this stage the central black hole is shrouded in dust. Later the dust clears out so that gas accretion around the black hole becomes visible and the galaxy enters a "variable-object phase."

A second team led by Seth Cohen, also of Arizona State University, observed 4,600 objects over many weeks and found that 1% of the objects fluctuated over time, indicating they contained central supermassive black holes feeding on gas from their surroundings and not shrouded in dust.

http://hubblesite.org/newscenter/newsdesk/archive/releases/2006/04/

http://arxiv.org/abs/astro-ph/0511423

Above: *This dazzling infrared image from NASA's Spitzer Space Telescope shows hundreds of thousands of stars crowded into the swirling core of our Milky Way galaxy. In visible-light pictures, this region cannot be seen at all because dust lying between Earth and the galactic center blocks the view. In this false-color picture, old and cool stars are blue, while dust features lit up by blazing hot, massive stars are shown in a reddish hue. Both bright and dark filamentary clouds can be seen, many of which harbor stellar nurseries. The plane of the Milky Way's flat disk is apparent as the main, horizontal band of clouds. The brightest white spot in the middle is the very center of the galaxy, which also marks the site of a supermassive black hole. Image courtesy NASA/JPL-Caltech/S. Stolovy (SSC/Caltech)*

10 January 2006
Spitzer Captures Our Galaxy's Bustling Center

A remarkable and spectacular infrared mosaic of the central region of our Milky Way has been acquired using images from the orbiting Spitzer Space Telescope. The new image reveals dramatic new details, from the star forming regions to hundreds of thousands of older stars.

"With Spitzer, we can peer right into the heart of our own galaxy and see breathtaking detail," said Susan Stolovy of the Spitzer Science Center at the California Institute of Technology in Pasadena. "This picture is crammed with fascinating features that we have just begun to explore."

This view has not been possible before on such a large scale. Dust in the plane of our Milky Way prevents visible light from reaching us from the central regions of the galaxy. Infrared light penetrates dust, allowing the newest infrared telescopes to see them. The high resolution possible from the vantage point of space improves the view from Spitzer.

The image reveals a region of space 900 light years across and covers an area about 2-degrees wide, or four full moon widths. Features within the new mosaic include a wide variety of dust clouds, including glowing filaments, wind-blown lobes flapping outward from the plane of the galaxy, and finger-like pillars.

"Our Spitzer data, combined with data obtained by other telescopes, will allow us to determine which of these objects are truly at the galactic center, and which are in spiral arms along the way," said Stolovy. "This survey will help us to better understand the mass distribution and structure of our own galaxy and how it compares to other galaxies."

(Editor: See the feature article by Michelle Thaller about the Spitzer Space Telescope elsewhere in this volume.)

http://www.spitzer.caltech.edu/Media/releases/ssc2006-02/index.shtml

11 January 2006
Hubble Telescope Reveals Thousands of Orion Nebula Stars

A large mosaic of the Orion Nebula presents the most detailed image yet of the 1,500 light year distant stellar nursery. In the spectacular image containing over a billion pixels, over a thousand new stars never seen before have been found: most are very faint and many are brown dwarfs.

The Orion Nebula is the nearest large star formation region visible to us. It is full of brilliant new stars, protostars surrounding by dusty disks, hot gas, cold dust and thousands

Below: *In one of the most detailed astronomical images ever produced, NASA's Hubble Space Telescope reveals the complex structure of the Orion Nebula. More than 3,000 stars of various sizes appear in this image – some never been seen before in visible light. The Orion Nebula is a picture book of star formation, from the massive, young stars that are shaping the nebula to the pillars of dense gas that may be the homes of budding stars. Image courtesy NASA,ESA, M. Robberto (Space Telescope Science Institute/ESA) and the Hubble Space Telescope Orion Treasury Project Team.*

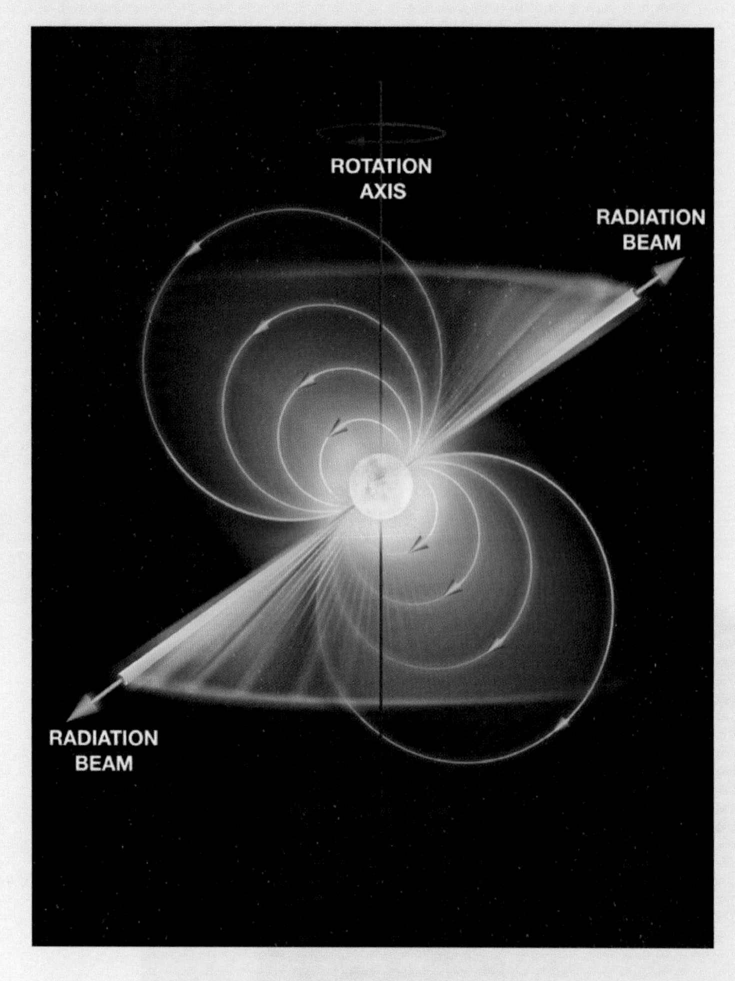

of stars. Since our Sun was possibly formed in a similar gas cloud, observing star formation in the Orion nebula at such close quarters represents an excellent opportunity. 105 Hubble orbits were required to complete the mosaic.

"The wealth of information in this Hubble survey, including seeing stars of all sizes in one dense place, provides an extraordinary opportunity to study star formation," said observation leader Massimo Robberto of the Space Telescope Science Institute, Baltimore.

http://hubblesite.org/newscenter/newsdesk/archive/releases/2006/01/

*■ **Above:** Illustration of a pulsar showing magnetic field lines and radio beams. Pulsars are spinning neutron stars that send out lighthouse-like beams of radio waves, or light, around as they spin. A neutron star is what is left after a massive star explodes at the end of its 'normal' life. With no nuclear fuel left to produce energy to offset the stellar remnant's weight, its material is compressed to extreme densities. The pressure squeezes together most of its protons and electrons to form neutrons; hence, the name 'neutron star'. Illustration courtesy Bill Saxton, NRAO/AUI/NSF.*

12 January 2006
Huge 'Superbubble' of Gas Blowing Out of Milky Way

A giant "superbubble" of hydrogen gas rising high above the plane the Milky Way galaxy has been detected by the Green Bank Radio Telescope (GBT). This unusual feature lies nearly 10,000 light years above the plane, and some energetic processes must have occurred to drive the bubble so high above the gravitational field of the galaxy.

Such high energies had to have come, astronomers argue, from supernova explosions and the intense stellar winds from an unseen cluster of young stars in one of our Galaxy's spiral arms.

"This giant gas bubble contains about a million times more mass than the Sun and the energy powering its outflow is equal to about 100 supernova explosions," said Yurii Pidopryhora, of the National Radio Astronomy Observatory (NRAO) and Ohio University.

The superbubble lies nearly 23,000 light-years from Earth. The astronomers discovered it by combining creating a mosaic of numerous smaller radio images along with ionized hydrogen images of the region.

"Finding this superbubble practically in our back yard is quite exciting, because these superbubbles are very important factors in how galaxies evolve," Jay Lockman of NRAO said.

http://www.nrao.edu/pr/2006/plume/

12 January 2006
Astronomers Discover Fastest-Spinning Pulsar

A fast spinning pulsar buried within the globular cluster, Terzan 5, has been discovered to be the fastest one on record. The pulsar, named PSR J1748-2446ad, lies 28,000 light-years from Earth in the constellation Sagittarius.

Its record setting pace was discovered by the Green Bank Telescope (GBT). The pulsar is spinning 716 times per second, easily beating the previous record of 642 Hz from a pulsar discovered in 1982. For reference, the fastest speeds of common kitchen blenders are 250-500 Hz.

"We believe that the matter in neutron stars is denser than an atomic nucleus, but it is unclear by how much. Our observations of such a rapidly rotating star set a hard upper limit on its size, and hence on how dense the star can be." said Jason Hessels, a graduate student at McGill University in Montreal.

The pulsar also has a companion star that orbits it once every 26 hours. The orientation of its orbit causes the pulsar to be eclipsed about 40 percent of the time. The long eclipse indicates the companion is probably a distorted shape, making it difficult to accurately measure their masses.

"If we could pin down these masses more precisely, we could then get a better limit on the size of the pulsar. That, in turn, would then give us a better figure for the true density inside the neutron star," explained Ingrid Stairs of the University of British Columbia. Such observational data is required to determine which theoretical models of internal nature of pulsars are correct.

http://www.nrao.edu/pr/2006/mspulsar/

http://arxiv.org/abs/astro-ph/0601337

19 January 2006

Dust Planetary Disks Around Two Nearby Stars Resemble Our Kuiper Belt

Hubble Space Telescope images have revealed two dusty disks around two stars that appear equivalent in size to our own Kuiper Belt. The Advanced Camera for Surveys was used in coronagraph mode, which produces the black central disk blocking out starlight, to acquire the images. The two stars, HD 53143 and HD 139664 are located about 60 light years from Earth and encircle the kinds of stars where there could be habitable zones capable of supporting life.

These new disks bring the total disks observable at visible wavelengths to nine, and are the oldest found optically so far. The first ever disk to be detected was around Beta Pictoris. The central area in both disks appear to be cleared out, suggesting planets probably exist within the systems.

The disk around HD 53143, a K-star slightly smaller than the Sun, is inclined obliquely to the line-of-sight and is about 1 billion years old. Its disk is about 55 astronomical units wide. By contrast, HD 139664, an F-star slightly larger than the Sun, is only 300 million years old. Its disk is seen almost edge on and the outer edge lies about 109 astronomical units from the parent star.

Just as the moons of Saturn and Uranus keep ring particles confined to discrete orbits with relatively sharp edges, so the presence of a clear

Below: *These false-color images taken with the Hubble Space Telescope's Advanced Camera for Surveys show two bright debris disks of ice and dust encircling stars around which there could be habitable zones and planets for life to develop. The disks seem to have a central area cleared of debris, perhaps by planets. The wide disk on the left surrounds HD 53143, a K-star slightly smaller than the Sun but about 1 billion years old. The narrow disk on the right encircles the star HD 139664, an F-star slightly larger than the Sun but only 300 million years old. The black central circle is produced by the camera's coronagraph which blocks the glare from the central star to allow the much fainter disks to be seen. Images courtesy NASA, ESA, and P. Kalas (University of California, Berkeley).*

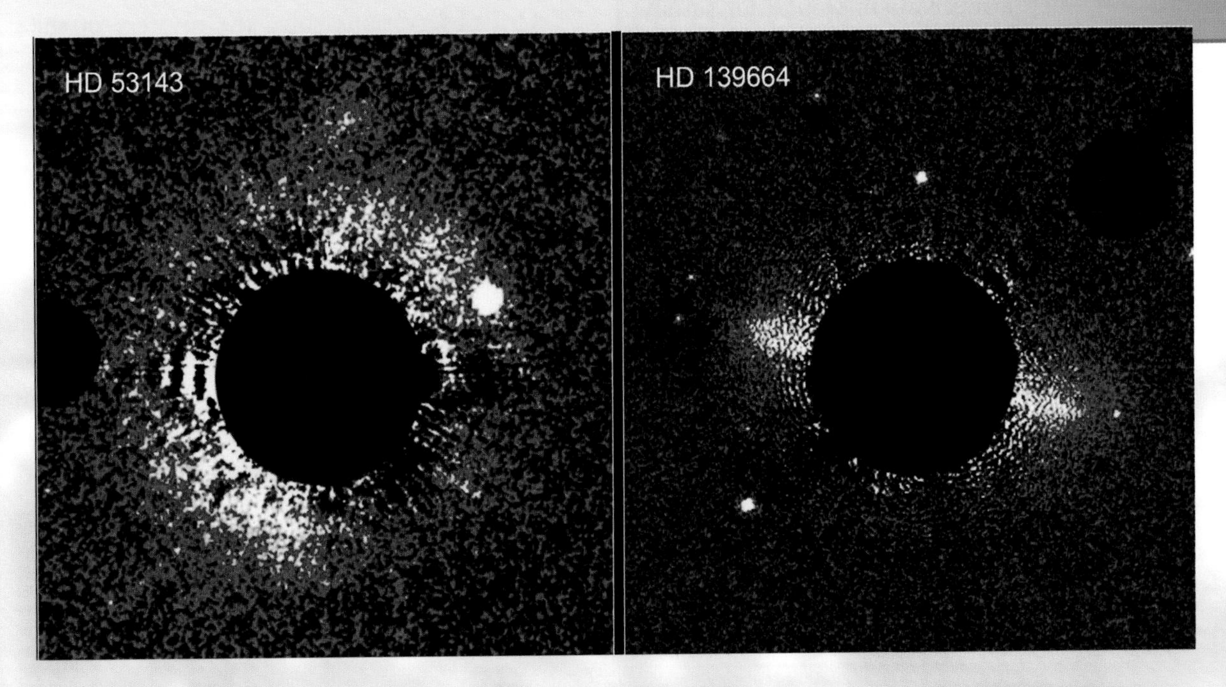

cutoff in these observed disks around the two stars indicate unseen planetary-type companions.

http://hubblesite.org/newscenter/newsdesk/archive/releases/2006/05/

http://arxiv.org/abs/astro-ph/0601488

30 January 2006

Most Milky Way Stars are Single

Common lore has been that most stars in the Milky Way are part of double or multiple systems.

Hence the news coming from the Harvard-Smithsonian Center for Astrophysics (CfA) is a little hard to believe. So here it is. Common wisdom is wrong. Most Milky Way stars are single.

This radical turnaround comes from the new understanding of the number of red dwarf stars in the galaxy. The common wisdom came from the fact that most bright stars, such as the Sun and visible stars in our night sky, do in fact form in multiple systems. However, recent studies have found that red dwarfs rarely occur in multiples, and because they are considerably more abundant than massive bright stars, the conclusion is that most stars in the Milky Way are single.

"By assembling these pieces of the puzzle, the picture that emerged was the complete opposite of what most astronomers have believed," said Charles Lada of CfA.

The findings have significant implications. The result suggests that most stars are born single. While some double systems may eject one star, such a mechanism would not satisfy the formation of the majority of stars. Also if the accepted wisdom that planetary systems are most likely to occur around single stars, this implies an abundance of planets. (See 25 January announcement).

http://www.cfa.harvard.edu/press/pr0611image.html

http://arxiv.org/abs/astro-ph/0601375

FEBRUARY 2006

3 February 2006

Detection of Hot Galactic Halo a Missing Link of Galaxy Formation

The quiescent disk galaxy, NGC 5746, located 100 million light years from Earth in the constellation of Virgo, is bathed in a massive halo of hot gas. Typically a feature of galaxy clusters that have visibly interacting members, detecting such a large halo in X-rays suggests this apparently stable galaxy is still accreting matter, probably through the gradual inflow of intergalactic gas.

The Chandra X-ray observatory found the halo extended 60,000 light years on either side NGC 5746.

"What we are likely witnessing here is the ongoing galaxy formation process," said Kristian Pedersen of the University of Copenhagen, Denmark.

Theory says spiral galaxies are formed from gigantic clouds of gas that collapse to form a spinning disk of stars and gas and dust. If this is the case, each spiral galaxy should be immersed in such haloes of hot gas. None had been seen except those linked to the outflow caused by a burst of star formation. This new result shows little star formation activity, and must be due to the gas infall.

Seen as a "missing link" of galaxy formation, astronomers are pleased to have found an example of the predicted halo at last.

"Our observations solve the mystery of the missing hot halos around spiral galaxies," said

Right: *Chandra observations of the massive spiral galaxy NGC 5746 revealed a large halo of hot gas (blue) surrounding the optical disk of the galaxy (white). The halo extends more than 60,000 light years on either side of the disk of the galaxy, which is viewed edge-on. Computer simulations and Chandra data show that the likely origin of the hot halo is the gradual inflow of intergalactic matter left over from the formation of the galaxy. Images courtesy NASA/CXC/Univ. Copenhagen / K.Pedersen et al. – X-ray; Palomar DSS – Optical.*

Above: *A team of astronomers from the UK, USA, Australia, Italy and Canada using the CSIRO Parkes radio telescope (pictured here) in eastern Australia has found a new kind of cosmic object: small, compressed 'neutron stars' that show no activity most of the time but once in a while spit out a single burst of radio waves. The new objects—dubbed Rotating Radio Transients or RRATs—are likely to be related to conventional radio pulsars (small stars that emit regular pulses of radio waves, up to hundreds of times a second). But the new objects probably far outnumber their old cousins, the scientists say. Image courtesy Shaun Amy.*

Pedersen. "The halos exist, but are so faint that an extremely sensitive telescope such as Chandra is needed to detect them."

http://chandra.harvard.edu/press/06_releases/press_020306.html

http://arxiv.org/abs/astro-ph/0511682

15 February 2006

Astronomers Find New Kind of Cosmic Object

A new kind of celestial object has been found by Australian astronomers observing with the Parkes radio telescope in New South Wales. The unusual objects are neutron stars, but unlike their active counterparts, the pulsar, these objects show no activity most of the time, and then suddenly turn on a single burst before returning to a quiescent state.

The new objects are called Rotating Radio Transients or RRATs. While many active pulsars, emitting hundreds of pulses a second, have been catalogued, astronomers observing these new objects suspect they far outnumber their active counterparts.

The bursts last from 2 to 30 milliseconds separated by minutes to hours of radio silence in the 11 new RRATs detected. Most of the sources have periods embedded within the short bursts, indicating they are rotating neutron stars.

"These things were very difficult to pin down," says CSIRO's Dr Dick Manchester, a member of the research team and a veteran pulsar hunter. "For each object we've been detecting radio emission for less than one second a day. And because these are single bursts, we've had to take great care to distinguish them from terrestrial radio interference."

Because these objects easily escape detection astronomers think there could be a few hundred thousand of them scattered throughout our galaxy alone.

http://www.atnf.csiro.au/news/press/images/rrats_pics/

http://xxx.lanl.gov/abs/astro-ph/0511587

21 February 2006

Spitzer Space Telescope Makes Hot Alien World the Closest Directly Detected

The Spitzer Space Telescope has detected the closest extrasolar planet to our solar system, and measured its surface temperature based on its infrared light output. The Jupiter-sized planet lies 63 light years away and orbits the star HD 189733 in the remarkably short period of 2.2 days. The planet's orbit is small, carrying it within 5 million kilometers of the star. Its surface temperature reaches over 1,500 degrees Fahrenheit (840 degrees C).

"This is the closest extrasolar planet to Earth that has ever been detected directly, and it presents the strongest heat emission ever seen from an exoplanet," said Drake Deming of NASA's Goddard Space Flight Center, Greenbelt, Md.

The planet disappears each orbit behind its parent star. The direct measurement was made by observing the infrared light of the star alone, and with the planet and star side by side. A simple subtraction reveals the infrared radiation from the planet alone.

http://www.spitzer.caltech.edu/Media/releases/ssc2006-07/release.shtml

http://xxx.lanl.gov/abs/astro-ph/0602443

rate of the universe. These types of supernovae all reach the same brilliance, and their dimness is a measure of their distance from us.

Another supernova in this galaxy recently made the news. SN 1979C has faded in visible light but it's still glowing brightly in X-rays, 25 years later.

(Editor: See 21 July 2005.)

http://www.eso.org/outreach/press-rel/pr-2006/phot-08-06.html

23 February 2006
VLT Image of SN 2006X in Spiral Galaxy Messier 100

Early in February two amateur astronomers, Shoji Suzuki and Marco Migliardi, discovered supernova 2006X in the spiral galaxy, M100. Detected at magnitude +17, it's turned out to be a type Ia supernova and was two weeks before its peak brightness. M100 is located 60 million light years from Earth in the constellation of Coma Berenices, and is easily visible in small telescopes.

Now the European Very Large Telescope has taken a highly detailed image of the galaxy, revealing its bright spiral arms, brilliant core and many HII star forming regions. Type Ia supernovae are important to astronomers because they are used to measure the distances to very distant galaxies, and hence the expansion

24 February 2006
Swift Satellite Detects Unusual Cosmic Explosion

A new kind of Gamma Ray Burst (GRB), whose characteristics are totally different to anything previously seen, was detected by the Swift spacecraft on 18 February. The object, called GRB 060218, appeared to be located 25 times closer and lasted 100 times longer than any previously known burst. The object was bright enough that amateur astronomers could easily view it directly with backyard telescopes. Dozens of telescopes around the world hurriedly re-arranged their observing schedules to observe the unusual event.

"The observations indicate that this is an incredibly rare glimpse of an initial gamma-rayburst at the beginning of a supernova," said Peter Brown, a graduate student at Pennsylvania State University and a member of the Swift science team.

__Above:__ This image of the type Ia supernova SN 2006X in the spiral galaxy M100 is a composite based on five images taken with FORS1 on Kueyen at the ESO Very Large Telescope by Dominique Naef, Eric Depagne and Chris Lidman (ESO). The final processing was done by Kristina Boneva, Haennes Heyer and Henri Boffin (ESO).

__Right:__ Scientists are studying a strange explosion that appeared on 18 February 2006, about 440 million light years away in the constellation Aries. The 'before' image on the left is from the Sloan Digital Sky Survey (SDSS). The 'after' image on the right is from Swift's Ultraviolet/Optical Telescope. The pinpoint of light from this star explosion outshines the entire host galaxy. Most other sources are foreground stars. Images courtesy SDSS (left), NASA/Swift/UVOT (right).

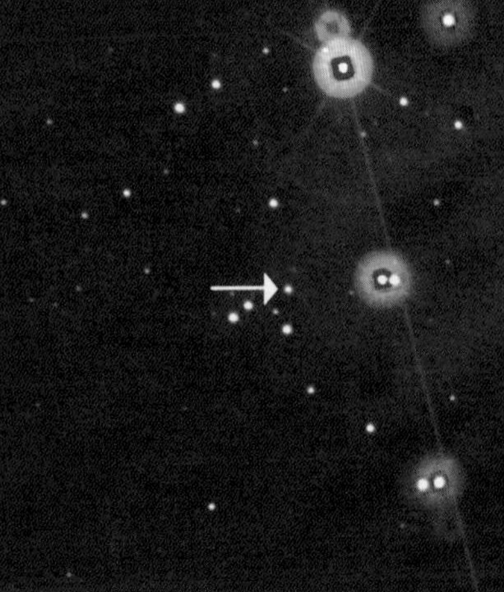

The burst lasted 2,000 seconds, instead of the typically tens of seconds, or in the case of short GRB's, milliseconds. The closest one ever recorded was in 1998, long before the Swift spacecraft was in orbit. Astronomers had waited a long time to catch one of these events relatively nearby. It occurred in a galaxy located 440 million light years away.

"This burst is totally new and unexpected," said Neil Gehrels, Swift principal investigator at NASA's Goddard Space Flight Center in Greenbelt, Maryland. "This is the type of unscripted event in our nearby universe that we hoped Swift could catch."

Early observations with the VLT in Chile revealed unusual spectral features.

"We expected to see the typical featureless spectrum of a gamma-ray burst afterglow, but instead we found a mixture between this and the more complex spectrum of a supernova similar to those generally observed weeks after the gamma-ray burst," said Nicola Masetti of the Italian Institute for Space Astrophysics and Cosmic Physics (IASF) in Bologna. "A supernova must be in the works."

Masetti speculates that this could be a supernova type Ic, a massive star exploding. If she is correct, astronomers have witnessed the very early stages of a supernova using some of the world's finest telescopes. Early detection is difficult when limited to optical light, but the fast acting Swift satellite opens the door to new breakthroughs in our understanding of these stellar eruptions. Use the Swift web site link for the latest news on this fascinating object.

http://swift.gsfc.nasa.gov/docs/swift/swiftsc.html

http://www.science.psu.edu/alert/Nousek2-2006.htm

http://www.nasa.gov/mission_pages/swift/bursts/oddball_burst.html

27 February 2006

Andromeda's Stellar Halo Shows Galaxy's Origin to be Similar to Milky Way

Measuring the rotation velocity of stars within the Andromeda Galaxy using the 10-meter Keck telescope, astronomers have concluded that our neighboring galaxy had a similar early history to that of our own Milky Way.

Over 10,000 stars were analyzed, and more than 1,000 belonged to a huge halo of stars that extends more than 500,000 light years from the galaxy's center. These stars show a low percentage of 'metals'. The expectations were that Andromeda's halo contained metal rich stars in contrast to the metal poor Milky Way halo. The measurements show that both galaxies must have evolved in similar ways for the first few billion years.

In the early universe galaxies formed around dark matter 'seeds' and in their early history merged with many small galaxy's. The stars in this outer halo would have formed early in the galaxies history and were not subjected to stellar

Below: *Giant galaxies weren't assembled in a day. Neither was this Hubble Space Telescope image of the face-on spiral galaxy M101. It is the largest and most detailed image of a spiral galaxy that has ever been released from Hubble. The galaxy's portrait is actually composed of 51 individual exposures taken with Hubble's Advanced Camera for Surveys and the Wide Field and Planetary Camera 2. Image courtesy NASA and ESA.*

evolution processes that build the percentage of metals in later generations of stars.

"Probably, both galaxies got started within a half billion years of the Big Bang, and over the next three to four billion years, both were building up in the same way by protogalactic fragments containing smaller groups of stars falling into the two dark-matter haloes," Scott Chapman of the California Institute of Technology explains.

http://pr.caltech.edu/media/Press_Releases/PR12801.html

http://www.astro.caltech.edu/~schapman/m31haloinfo.html

http://xxx.lanl.gov/abs/astro-ph/0602604

28 February 2006

Giant Galaxy Portrait of M101 by the Hubble Space Telescope

The largest Hubble telescope image released to date is of the huge pinwheel shaped galaxy, M101. A total of 51 images went into the final image, measuring 16,000 by 12,000 pixels.

The fine resolution of the image, the most detailed ever of a nearby galaxy, has revealed over 3,000 previously undetected star clusters. The galaxy lies about 25 million light years away in the constellation of Ursa Major, and spans nearly double the diameter of the Milky Way. Estimates give the stellar population of about one trillion stars.

Images that make up this image were taken over a ten year period, from 1994 to 2003, and include date from the Wide Field and Planetary Camera 2 (WFPC2) and the Advanced Camera for Surveys (ACS).

Large numbers of galaxies lie in the background, giving the image a unique depth perspective. One barred spiral in particular, probably of similar size to M101, is so far away it appears tiny in the image. Rich tendril-like dust clouds track along the spiral arms of M101 almost to the bright central core of the galaxy.

http://www.spacetelescope.org/news/html/heic0602.html

http://hubblesite.org/newscenter/newsdesk/archive/releases/2006/10/

MARCH 2006

2 March 2006

A Shocking Surprise in Stephan's Quintet

The well known cluster, Stephan's Quintet, is viewed in a new light by a combination of images from the Spitzer Space Telescope and the Calar Alto Observatory in Spain. The cluster, located 300 million light years away, reveals a dramatic shock wave, the largest ever seen, produced by the collision of hot gas between two merging galaxies. With the approach speed of 1.6 million kilometers per hour, the shock wave extends over a region of space larger than our own galaxy.

***Right:** This false-color composite image of the Stephan's Quintet galaxy cluster clearly shows one of the largest shock waves ever seen (green arc), produced by one galaxy falling toward another at over a million miles per hour. Four of the five galaxies in this image are involved in a violent collision. One galaxy, the large spiral at the bottom left of the image, is a foreground object and is not associated with the cluster. This image uses data from the Infrared Array Camera on NASA's Spitzer Space Telescope and data in visible light from the Calar Alto Observatory in Spain. Image courtesy NASA/JPL-Caltech/Max-Planck Institute/P. Appleton (SSC/Caltech)*

The arc was detected from the ground, and follow-up observations from the Spitzer Space telescope zeroed in on the hot gas in the shock wave. Spitzer found it to be made of very turbulent hydrogen gas. NGC 7218b, the left hand object of two small bright regions in the center of the image, is the culprit. The other three galaxies in the group have already stripped most of the hydrogen gas from the cluster. The spiral galaxy at the lower left is a foreground object.

http://xxx.lanl.gov/abs/astro-ph/0602554

http://www.spitzer.caltech.edu/Media/releases/ssc2006-08/index.shtml

8 March 2006
Scientists Piece Together the Most Distant Cosmic Explosion

The widely observed Gamma-Ray Burst, GRB 050904, was the result of the collapse of a massive star to form a black hole, astronomers have deduced. The burst that occurred in September 2005 was the farthest every observed, located at a redshift of 6.3, placing it within a million light years of the edge of the observable universe.

"This was a massive star that lived fast and died young," said David Burrows, of Penn State University. "This star was probably quite different from the kind we see today, the type that only could have existed in the early universe."

Astronomers have observed very few objects this far from Earth. Only one quasar has been detected at a higher redshift. In future, exploding stars from one of the first generations of stars to have formed after the Big Bang could be detected, making this and other GRB's a unique opportunity to study the universe at a location typically beyond the range of most telescopes.

"Because the burst was brighter than a billion suns, many telescopes could study it even from such a huge distance," added Burrows.

The orbiting Swift observatory detected the burst. Unusual features of this GRB include the length of the burst, lasting about 500 seconds, along with multiple flares, indicating that the newly formed black hole was not formed instantly. Evidence of time-dilation as also present and hints of heavy element formation were observed.

"We designed Swift to look for faint bursts coming from the edge of the universe," said Neil Gehrels of NASA Goddard Space Flight Center in Greenbelt, Maryland, Swift's principal investigator. "Now we've got one and it's fascinating. For the first time we can learn about individual stars from near the beginning of time. There are surely many more out there."

http://www.science.psu.edu/alert/Burrows3-2006.htm

http://xxx.lanl.gov/abs/astro-ph/0509660

http://xxx.lanl.gov/abs/astro-ph/0509697

15 March 2006
The Cosmic Dance of Distant Galaxies

The multi-object GIRAFFE spectrograph on the Very Large Telescope (European Southern Observatory, ESO) has revealed inner details of distant galaxies. The large 8.2-m telescope has revealed that these distant galaxies had the same amount of dark matter 6 billion years ago as the do today.

"Dark matter, which composes about 25% of the Universe, is a simple word to describe something we really do not understand," said Hector Flores of the Paris Observatory. "From looking at how galaxy rotates, we know that dark matter must be present, as otherwise these gigantic structures would just dissolve."

Below: *The new results obtained with the multi-object GIRAFFE spectrograph on the ESO Very Large Telescope seem to show that collisions and mergers are important in the formation and evolution of galaxies. Here, such a collision is shown in this artist's impression. Illustration courtesy ESO.*

In the past, telescopes could only collect enough light from distant galaxies to determine their overall spectrum. Their great distance made them too small for spectrographs to resolve any finer detail. In order to measure the rotation of galaxies at different distances from their centers, the new spectrograph employs multiple micro-lenses that feed light from different parts of the distant galaxies into the spectrograph.

"GIRAFFE on ESO's VLT is the only instrument in the world that is able to analyze simultaneously the light coming from 15 galaxies covering a field of view almost as large as the full moon," said Mathieu Puech of the Paris Observatory. "Every galaxy observed in this mode is split into continuous smaller areas where spectra are obtained at the same time."

Astronomers found that 40% of the galaxies showed evidence of disruption by collisions. The remainder revealed rotation curves that matched those seen in present day, nearby galaxies, indicating the amount of dark matter is the same at both epochs.

http://www.eso.org/outreach/press-rel/pr-2006/pr-10-06.html

http://arxiv.org/abs/astro-ph/0603562

http://arxiv.org/abs/astro-ph/0603563

15 March 2006

Astronomers Discover a River of Stars Streaming Across the Northern Sky

A huge stream of stars is being tidally stripped from a globular cluster of stars called NGC 5466, and spans over 45 degrees in our sky. The newly discovered stream spans from a point south of the Big Dipper almost to Arcturus, in Boötes.

Existence of the stream was smothered by foreground stars, and it took a detailed study of data in the Sloan Digital Sky Survey to reveal the tidal streaming, which spreads both in front and behind the cluster. Last year astronomers had found evidence for a tidal stream that was 4 degrees long. This new study traces it to over ten times as long.

Below: Time-line of the universe. The expansion of the universe over most of its history since the Big Bang has been relatively gradual. The notion that a rapid period of 'inflation' preceded this more gradual expansion was first put forward 25 years ago. New WMAP observations favor specific inflation scenarios over other long held ideas. Image courtesy NASA/WMAP Science Team.

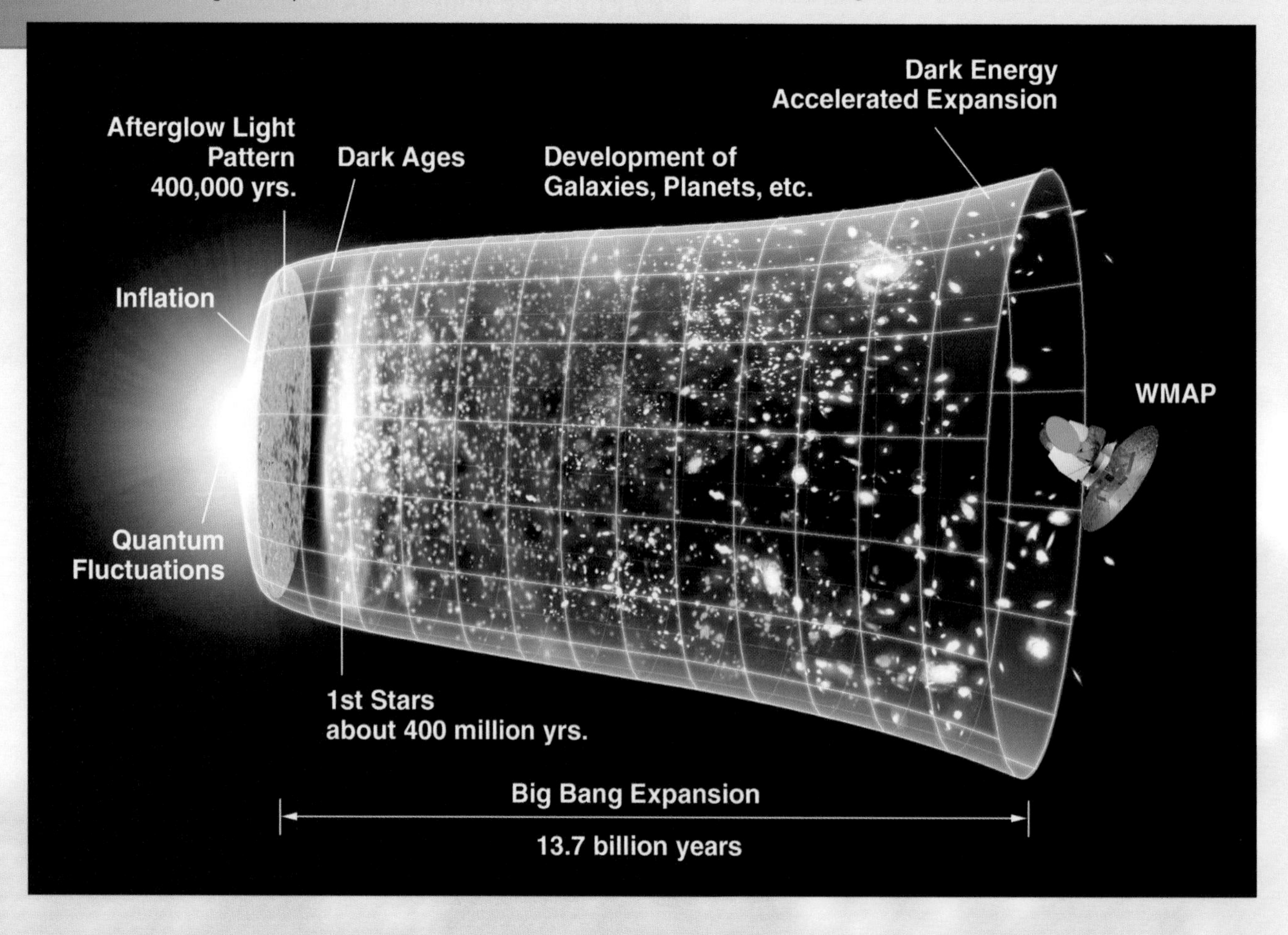

Detecting which stars are members of the stream required some sleuthing. The clue came from the type of stars found on globular clusters.

"It turns out that, because they were all born at the same time and are situated at roughly the same distance, the stars in globular clusters have a fairly unique signature when you look at how their colors and brightnesses are distributed," says Grillmair.

By labeling those stars that had a high probability of previously being a member of NGC 5466, evidence of the streaming became obvious. About 350 members were found that arc high over the disk of the Milky Way. They lie about 76,000 light years away. Tidal tails are known to exist near at least 40 other globular clusters.

Tidal streaming of stars allows detailed structure of the gravitational attraction of different parts of our galaxy, essentially revealing the hills and valleys of gravitational potential in our galaxy.

http://www.sdss.org/sdss.html

http://pr.caltech.edu/media/Press_Releases/PR12811.html

http://arxiv.org/abs/astro-ph/0602602

16 March 2006

New Satellite Data on Universe's First Trillionth of a Second

The Wilkinson Microwave Anisotropy Probe (WMAP) has already tied the age of the Universe to 13.7 billion year old, and now, three years later, has provided substantial support for the theory of inflation. The evidence lies embedded in the microscopic temperature variations in the Cosmic Microwave Background (CMB) radiation that WMAP has measured to such high precision.

"It amazes me that we can say anything about what transpired within the first trillionth of a second of the universe, but we can," said Charles Bennett of The Johns Hopkins University and lead investigator of the WMAP mission. "We have never before been able to understand the infant universe with such precision. It appears that the infant universe had the kind of growth spurt that would alarm any mom or dad."

Three years of data from WMAP has revealed that a key prediction of inflation – that the brightness of large temperature fluctuations would be greater than smaller ones – is indeed borne out by the data. It confirms the idea that in the very first fraction of a second after the Big Bang, the universe underwent exponential growth in size, and the result of that expansion is imprinted forever on the CMB.

Above: *This image composite compares a visible-light view (left) of the so-called 'Cigar galaxy', also called M82, with an infrared view (right) from NASA's Spitzer Space Telescope of the same galaxy. The visible-light picture shows only a bar of light against a dark patch of space. Longer exposures of the galaxy (not shown here) have revealed cone-shaped clouds of hot gas above and below the galaxy's plane. The Spitzer data show that the galaxy is also surrounded by a huge, hidden halo of dust (red in infrared image). The dust particles (red) are being blown out into space by the galaxy's hot stars (blue). Images courtesy NOAO – Visible; NASA/JPL-Caltech/C. Engelbracht (University of Arizona) – Infrared.*

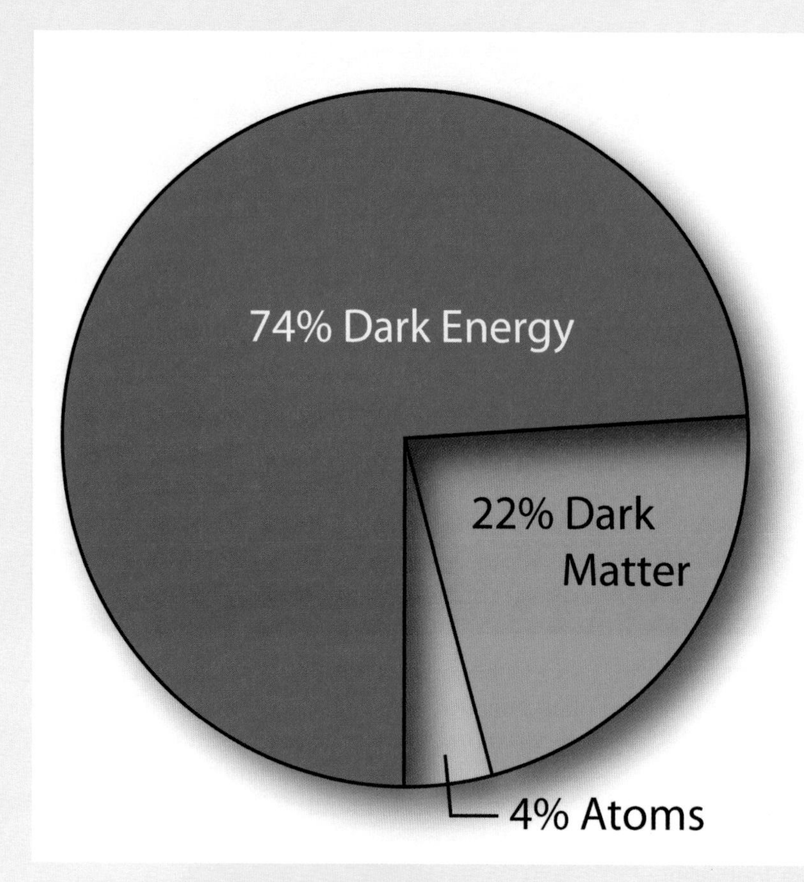

WMAP measured the subtle and long sought after polarization of the CMB across the entire sky. It is the weakest signal ever detected, and is one hundredth of the level of temperature variations found in the CMB itself. The new results provide information for astronomers to determine which of the many inflation scenarios is the most likely one to have occurred. During this period the conditions were set for the formation of the first stars a few hundred million years later.

"Inflation was an amazing concept when it was first proposed 25 years ago, and now we can support it with real data," said Gary Hinshaw of the Goddard Space Flight Center in Greenbelt, Maryland.

http://www.nasa.gov/vision/universe/starsgalaxies/wmap_pol.html

http://arxiv.org/abs/astro-ph/0603450

http://arxiv.org/abs/astro-ph/0603449

Above: *WMAP data reveal that the contents of the universe include just 4% atoms, the building blocks of stars and planets. Dark matter comprises 22% of the universe. This matter, different from atoms, does not emit or absorb light. It has only been detected indirectly by its gravity. 74% of the Universe, is composed of 'dark energy', that acts as a sort of an anti-gravity. This energy, distinct from dark matter, is responsible for the present-day acceleration of the universal expansion. Image courtesy NASA/WMAP Science Team.*

16 March 2006
Spitzer Space Telescope Finds Billowing Dust from M82

Huge volumes of dust are being ejected from the well known galaxy, M82, from regions undergoing rapid bursts of star formation. New images from the Spitzer Space Telescope (SST) reveal the spectacular red clouds of dust extending either side of the cigar-shaped galaxy like the wings of a butterfly.

"We've never seen anything like this," said Charles Engelbracht of the University of Arizona, Tucson. "This unusual galaxy has ejected an enormous amount of dust to cover itself with a cloud brighter than any we've seen around other galaxies."

The SST also found that the dust carries significant quantities of polycyclic aromatic hydrocarbons, compounds similar to those found in a smoking barbecue. The remarkable finding is that these clouds extend 20,000 light years away from the plane of the galaxy, far bigger than any found before. Previously seen cone-shaped clouds of hot gas suggested that most of the energy was coming from massive stars in the central regions of the galaxy. The lack of the cone extending far into space suggests that stars all over the galaxy are taking part in the massive ejections of dust.

"Spitzer showed us a dust halo all around this galaxy," said Engelbracht. "We still don't understand why the dust is all over the place and not cone-shaped."

M82 is undergoing interaction with its nearby spiral neighbor, M81, and this interaction is driving the massive star formation in M82. M82 is a bright galaxy, visible in backyard telescopes, and lies about 12 million light years away in Ursa Major.

http://www.spitzer.caltech.edu/Media/releases/ssc2006-09/release.shtml

http://arxiv.org/abs/astro-ph/0603551

22 March 2006

A Unique 'Brown Dwarf' is Discovered Right in our Solar Neighborhood.

A nearby star to the Sun has been found to harbor a brown dwarf, making it the second closest ever discovered. The star, located in the constellation of Pavo, was discovered quite recently, and lies 12.7 light years from the Sun. Its low mass, less than one-tenth the mass of our Sun, makes it a very low luminosity object and consequently hard to find. The new brown dwarf orbits about more than 400 million miles from its dim parent star. Most brown dwarfs have been found around stars at least half a solar mass, making the finding even more surprising. The new object is called SCR 1845-6357B.

The mass of the brown dwarf is between nine and sixty-five times the mass of Jupiter. (See 15 March 2006 for another brown dwarf mass determination)

"Besides being extremely close to Earth and in orbit around a very low-mass star, this object is a 'T dwarf ' - a very cool brown dwarf with a temperature of about 750 degrees Celsius (1,382 degrees Fahrenheit)," said Beth Biller, a graduate student at The University of Arizona.

The brown dwarf is a member of the 25 closest star systems to the Sun, all known for decades, but only recently have the nearby brown dwarfs been discovered. There may be more out there yet to be found.

"This newly found brown dwarf is a valuable object because its distance is well known, allowing us to determine with precision its intrinsic brightness", said Markus Kasper of the European Southern Observatory. "Moreover, from its orbital motion, we should be able in a few years to estimate its mass. These properties are vital for understanding the nature of brown dwarfs."

If more T-dwarfs in binary systems are found in the solar neighborhood, this would have significant implications for theories of formation of brown dwarfs that suggest single objects would be formed more readily than binary ones. A closer T-dwarf orbits Epsilon Indi (in the constellation of Indus) only 11.8 light years away. Both objects were discovered using the ESO Very Large Telescope with an adaptive optics system.

http://www.eso.org/outreach/press-rel/pr-2006/pr-11-06.html

http://arxiv.org/abs/astro-ph/0601440

http://arxiv.org/abs/astro-ph/0309256

An artist's impression of the SCR 1845-6357 stellar system – a new brown dwarf in the solar neighbourhood The small red star is shown in the background while the newly discovered brown dwarf is at front. Illustration courtesy ESO.

2

Astronomers Meet to Discuss the MYSTERIES of the UNIVERSE ...

... Another night spent looking into the question of Black Holes

In January 2006, over 3,100 astronomers, educators, and journalists crowded into the conference hotel in Washington, D.C. for what became the largest gathering of astronomers ever held. *Martin Ratcliffe* was there, and here he picks out the highlights of that meeting.

. . . The 207th American ASTRONOMICAL SOCIETY Meeting

THE RECORD gathering at the 207th American Astronomical Society (AAS) Meeting was larger than meetings of the International Astronomical Union, although the record may not stand for long.

The strength of the attendance at the meeting reflected the vigorous growth of astronomical research over the past decade. The AAS meet twice per year, in early January and in May/June. The winter meetings are always larger, and have typically reached 1,800 to 2,500 in recent years. The meetings in Washington DC, which occur every four years and are rapidly becoming gatherings of Olympic proportions, are the largest, due in some part to the proximity of NASA Headquarters, the Goddard Space Flight Center and numerous large universities up and down the East Coast with active astronomical research programs.

To review everything at the meeting would take volumes, so here's a brief taste of a few of the highlights.

The meetings are a mixture of concurrent sessions with 15-minute research papers and longer plenary sessions that give in-depth overviews of current active areas of research. These invited review talks are the highlight of each meeting. They give an up to date survey of the most current results of recent active research on the cutting edge of astronomy. To give a taste of the breadth of the topics covered, here is a list of this year's invited talks:

- James Gunn ***(Princeton)*** – Recipient of the Henry Norris Russell Lectureship for a lifetime of achievement in astronomy
- **Detecting Gravitational Waves with interferometers** by Nergis Mavalvala (M.I.T.); (see the article abut LIGO by Mavalvala's colleague, Laura Cadonati)
- **Studying Stellar Evolution with Seismology** by Sarbani Basu (Yale)
- **Massive Stars and Feedback** by Sally Oey, (U. Michigan)
- **High Energy Cosmic Rays** by James W. Cronin (U. Chicago)
- **The Vision for Space Exploration** by Michael Griffin (NASA)
- **Measuring Cosmological Parameters** by Wendy Freedman (Carnegie Observatories)
- **Titan: A Fiercely Frozen Echo of the Early Earth,** by Tobias Owen *(U. Hawaii)*
- **NSF's Role in the Future of Astronomy** *by* Michael Turner *(NSF)*
- **Using the Tools of Science to Teach Science** by Carl E. Weiman *(U. Colorado)*
- **Cosmic Microwave Background Radiation** by George Efstathiou *(Inst. of Astronomy, Cambridge, UK)*
- **Large Scale Simulation of the Galaxy/AGN population** by Simon White *(MPI fur Astrophysik)*
- **Supernova Gamma-Ray Burst Connection** by Stan Woosley *(UCSC)*
- **Black Hole Astrophysics in the New Century** by Christopher Reynolds *(U. Maryland)*
- **Submillimeter Selected Galaxies** *by* Andrew Blain *(Caltech)*.

In addition to verbal presentations, sometimes given by doctorate students (90 at this meeting) facing an intimidating barrage of questions from experts in their respective fields, many astronomers decide to present their research via poster papers in the display area.

A dizzying array of poster papers are displayed each day, covering a wide range of topics. On the first day you'll find among the thirty or so topics; the newest information relating to Gamma Ray Bursts, Interacting binary stars, Circumstellar disks, Astrobiology, Evolution of galaxies, and Probing and understanding effective learning and teaching .

The next day the topics change. This time there's Astronomy in the K-12 classrooms, Extrasolar planets: Current searches, Stellar evolution, Neutron stars and x-ray binaries, Early results from Swift, and the newest Spitzer Space Telescope Observations.

You also hear about future planned instruments, both ground-based instruments like the Large Synoptic Survey Telescope, the Allen Array, and space-based missions like Constellation-X – a next generation X-ray telescope comprised of four separate telescopes – or GLAST, a large area Gamma-ray telescope. These presentations give science writers and astronomers insight into the dramatic discoveries likely to be discussed at future AAS meetings.

The exhibit halls include some major display from some of NASA's flagship missions, such as the Chandra X-ray Observatory and the Spitzer Space Telescope, as well as details of proposed missions as yet unfunded or in development.

Images from Top to Bottom

■ **Robert Nemiroff** *and* **David Band** *discuss the poster paper about the COMPLETE survey of star formation regions. Another poster paper in the background provides details of a study of M51.*

■ **Rahul Sheff** *(U. Maryland) displays his research poster paper about theoretical modeling of spur formation in spiral galaxies.*

■ **Blair Reaser** *(Swarthmore College) displays a spectacular montage of the Small Magellanic Cloud as part of an emission line survey made of the LMC/SMC using the Curtis Schmidt Telescope at the Cerro Tololo InterAmerican Observatory.*

■ **Jill Tarter** *and* **Peter Backus** *of the SETI Institute discussed the Allen Telescope Array with conference delegates. The Array is being built at the existing Hat Creek Observatory, run by the Radio Astronomy Laboratory at Berkeley, and located in the Cascades just north of Lassen Peak (California).*

■ *The Chandra X-ray Observatory's display in the AAS exhibit hall is a frequent rendezvous place for discussions and debate about current and future observing plans, in addition to visually stunning imagery and free literature.*
All images courtesy Martin Ratcliffe.

The general public get to hear about news from an AAS meeting through the conduit of the press room. Dr Steven Maran, AAS Press Officer, has steered this process for many years, and is now assisted by Dr Lynn Cominsky of Sonoma State University. With thousands of technical papers pouring into the AAS offices prior to the meeting, Maran and associates scour the proposed papers for some of the top news items that are likely to make the front page of the New York Times, Washington Post, and the wide variety of on-line news sources. The large quantity of research being presented at the meetings is distilled into manageable blocks, giving journalists the time to concentrate on a focussed selection of topics that are considered most likely to interest the general public.

When Maran started as Press Officer at the 1985 Tucson meeting, "we had about five reporters at a January meeting and less at a June meeting", he said. "We peaked at a meeting in DC about 10 yrs ago, with 204 registered press," although about 100 is more typical today.

The press room is a hive of activity for the entire conference. In addition to the featured speakers, hundreds of concurrent paper sessions, poster papers, and the vast exhibit area, there is a full slate of press conferences every day. At each press conference one of the top stories is presented by a panel of researchers, and with an independent commentator as well. This gives the science journalist a unique opportunity to ask questions and develop a story that will appear in tomorrow's newspapers. Other journalists in the press room are on longer deadlines, such as those researching for a book, for a television series, or simply keeping abreast of current research.

Top: *Well-known science writer,* ***Govert Shilling****, and AAS Press Officer,* ***Steve Maran****, work in the press room.*

Above Left: Robert Lupton *(left, Princeton) is interviewed by Dutch science journalist and author,* ***Govert Schilling****, about a Sloan Digital Sky Survey discovery of a dwarf galaxy currently merging with the Milky Way.*

Above Right: *A press conference announcing a variety of discoveries about double stars (Polaris) and Type Ia supernovae. Those seated are (left to right)* ***Armin Rest*** *and* ***Nick Suntzeff*** *(Cerro Tololo InterAmerican Observatory),* ***Steven Howell*** *(WIYN Observatory),* ***Edward Sion*** *(Villanova U.), and* ***Nancy Evans*** *(Center for Astrophysics, Harvard U.).*

All images courtesy Martin Ratcliffe.

One of the highlights of a January AAS meeting is a presentation by the incumbent NASA Administrator, a post currently held by Dr. Michael Griffin. With a new budget due to be published shortly after the AAS meeting, interest was high for Griffin's presentation. Nothing gets astronomers attention like reduced money for basic science, and nothing in Griffin's speech gave many astronomers much satisfaction. He explained his strategy for returning America to the Moon, a task that is complete with an underlying national strategic decision to have easy access to space using safe launchers. There was dark news for education and public outreach professionals. "NASA is not the Department of Education," said Griffin, a comment that had a chilling effect on the highly successful network of education and public outreach educators in the room. Clearly either private funding from independent sources or from the Department of Education or Department of Energy will have to replace whatever is lost from NASA's budget if the benefits of the growth of the last decade are to be maintained. Someone must be touting NASA successes to willing students in schools and to the general tax-paying public other than the 30-second sound bite on CNN.

An emotional highlight of Griffin's speech was the posthumous presentation of NASA's Exceptional Scientific Achievement medal to John Bachall. It was presented to Neta Bachall in recognition of her late husband's lifetime achievement in astronomy. Neta and her daughter Orli were present and received the award to an overwhelming and emotional standing ovation.

The presentation read "In recognition for his extraordinary service to NASA's space astronomy program throughout his career, including his leadership of the National Research Council's Decadel Survey and tireless advocacy for the Hubble Space Telescope".

Many smaller meetings take place during the conference. One was a celebratory cake for the 50th anniversary of the National Radio Astronomy Observatory. NRAO President, Dr. Fred Lo, used the occasion to present an award to Acree Chung. She was the prizewinner in a special competition to create the most visually interesting presentation of NRAO images. Numerous news worthy radio astronomy news releases occurred during the conference and the highlights were presented in a special news conference during the week.

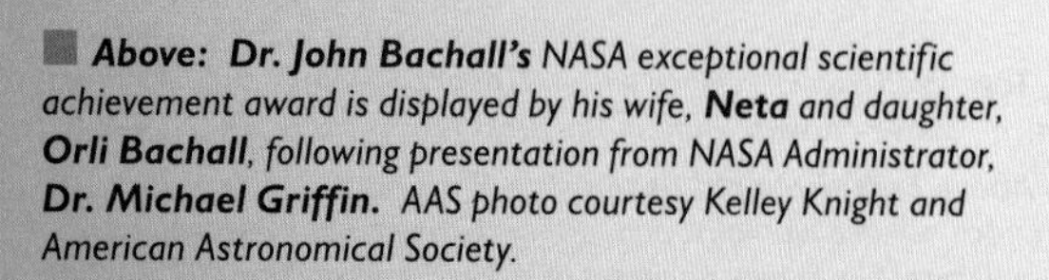

Above: Dr. John Bachall's *NASA exceptional scientific achievement award is displayed by his wife,* **Neta** *and daughter,* **Orli Bachall**, *following presentation from NASA Administrator,* **Dr. Michael Griffin.** *AAS photo courtesy Kelley Knight and American Astronomical Society.*

Above - inset: *The NASA exceptional scientific achievement award and medal presented to Neta Bachall on behalf of her late husband,* **Dr. John Bachall.** *AAS photo courtesy Kelley Knight and American Astronomical Society.*

Right: *Radio Astronomy Image contest winner,* **Acree Chung**, *with her first prize award for the Virgo A composite image. Image courtesy Martin Ratcliffe.*

A first for the AAS was the astronomical visualization session organized by Dr. Frank Summers. This session was a surprise to AAS organizers, since it was first planned for a small room, but as interest grew prior to the conference the room was switched to a larger venue – even so, it was still standing room only. Astrophysical simulations play a large part in research, and the large number of simulations, many being transferred to multiple channels simultaneously for the unique perspective of a hemispherical planetarium dome, speak to the wide audience such simulations reach. This session is sure to be the first of many.

AAS meetings are often the venue for the unveiling of the most recent entry into the "largest image ever made" category of astronomical images. Two images, the Orion Nebula taken by the Hubble Space Telescope, and the galactic center taken by the Spitzer Space Telescope, made the headlines this time around.

Top: *A special press conference reviewing astronomical results celebrating the 50th anniversary of the National Radio Astronomy Observatories was presented by, (from left to right)* ***Scott Ransom*** *(NRAO),* ***Chris Carilli*** *(NRAO),* ***Crystal Brogan*** *(U. Hawaii),* ***James Ulvestad, Mark Reid*** *and* ***Dale Frail*** *(all NRAO).*
AAS photo courtesy Kelley Knight and American Astronomical Society.

Above: ***Massimo Roberto*** *poses next to a giant print of the Orion Nebula. The image presents 33,000 x 36,000 pixels and required 104 orbits of the Hubble Space Telescope to acquire it. It is the largest mosaic of a star forming region made to date by HST. The image was unveiled at the Washington D.C. AAS meeting on 11 January 2006. Image courtesy Martin Ratcliffe.*

Left: *The new Spitzer Space Telescope mosaic of the Galactic Center is displayed behind astronomers* ***Farhad Yusef-Zadeh*** *(Northwestern U.) and* ***Susan Stolovy*** *(Spitzer Science Center). Image courtesy Martin Ratcliffe.*

Some astronomers use press conferences to provide excellent visual aids to describe phenomena observed in the universe. Perhaps the most innovative was the use of a blueberry bagel and flashlight by Rogier Windhorst and Amber Straughn to illustrate why orientation of accretion disks to our line-of-sight can block brilliant light emitted by material that is about to fall into a black hole.

The newest observatory to come on line is the LIGO gravitational wave interferometer. Nergis Malvalvala provided unique insight into the expected discoveries to come from LIGO, and Mike Turner, of the National Science Foundation that provided funding for LIGO, was on hand to provide additional commentary.

Hundreds of other presentations constitute the most vibrant and dynamic meeting of astronomers. By the second day of the meeting, most attendees agree that digesting as many of each day's talks as possible is rather like drinking from a fire hose. By the third day, brains are tired and overtaxed, and by Thursday, the last day of the conference, delegates begin filing to airports, train stations, and cars for the journey home. A familiar sight in many connecting airports across the country is a graduate student carrying a long cardboard tube containing his or her poster paper.

New friends and acquaintances have been made. New observing plans have been carved out. Plans for future instrumentation have been drawn on numerous restaurant napkins. Each year new discoveries are made, and presented. They are discussed, debated, and challenged by scientists eager to make their mark on our quest to understand the universe. Some ideas are tossed on the garbage heap - others become the fodder of new textbooks. AAS meetings are a place where the real human elements of scientific endeavor are laid bare. Careers can be made, funding can be won or lost, and universities clamor for attention. And yet, out of this human dynamic, the exploration of the universe inches forward, and the history of science is being written.

Future meetings of the American Astronomical Society can be found on the Society's web site, **http://www.aas.org.**

Images from Top to Bottom

*Astronomers **Rogier Windhorst** (left) and **Amber Straughn** (ASU) use a bagel and flashlight to creatively explain why a dense accretion disk around a black hole allows light from the hot gas descending into the black hole to be viewed only from certain orientations of the disk.*

***Nergis Malvalvala** (MIT) and **Michael Turner** (NSF) lead a Press Breakfast about LIGO on the opening day of the AAS meeting.*

*Scientists pack into a room to hear the latest results from the SWIFT Gamma-Ray observatory from its principle investigator, **Neil Gehrels**.*

All images courtesy Martin Ratcliffe.

A wide range of activities occur at meetings of the American Astronomical Society. Plenary sessions, exhibits and extensive poster sessions, special prizes for significant research, press conferences, special meetings on specific topics, and opportunities to catch up on the latest research, all make AAS meetings some of the most important events on the astronomical calendar.

All images courtesy Martin Ratcliffe.

3
The STATE of the UNIVERSE...

In this article *James Kaler*, Professor Emeritus of Astronomy, University of Illinois highlights the most significant discoveries of the year 2005-06, and places them in context of our broader astronomical knowledge.

...What we think WE KNOW and what WE DON'T

SEVERAL YEARS ago I bet an editor of a national astronomy magazine that the then-new discoveries about the overall nature of the Universe – described below – would not hold up. I lost (payment still due). They did hold up, discoveries that gratified cosmologists while also provoking them with awesome new challenges and uncertainties. Within this framework, that of the Universe at large, we have made major inroads into understanding its components: galaxies, their clusters, and their stars. And from the stars pour discoveries of other planets and planetary systems that lead right back to Earth, giving us a sense of our place within the structure of the Whole.

The Foundation

Cosmology, the study of the cosmos at large, has been with us since the dawn of humanity. We have a deep need to know our origins and our fate. One could argue that as a science it began with Edwin Hubble's eighty-year-old revelation that the then-mysterious 'spiral nebulae' were actually other galaxies. He followed this by showing that the redshifts in the spectra of relatively nearby galaxies (which imply recession) correlate nicely with the galaxies' distances. The Universe is expanding, the separation between galaxies (those not bound to each other by gravity) increasing, a concept that fitted right into Einstein's theory of general relativity, which was in turn necessary to understand it.

Hubble's original redshifts, which were actually measured by Vesto M. Slipher, were tiny, just 0.003 (that is, the wavelengths of spectrum lines were shifted to the red end of the spectrum by 0.3 percent), the greatest speeds about 1000 kilometers per second. The observations that pertain to the modern Hubble diagram now extend to a redshift of 1.7 (a 170 percent shift) which – invoking relativity - leads to a 'look-back time' of 9 billion years ago. Over the eight decades since Hubble's revelation, astronomers

Above: *The Hubble Space Telescope, focusing for 11.3 days on an area in Fornax just three arcminutes wide, revealed 10,000 galaxies tumbling to billions of light years away, each with 100 billion or more stars. And these are just the big ones. The more distant they are, the faster they recede from us. This is our Universe, this is our home. Image courtesy NASA, ESA, S. Beckwith (STScI) and the HUDF Team.*

in the Virgo cluster and beyond, which allows us to obtain the galaxies' distances. (Cepheids are variable supergiant stars whose oscillation periods are strictly related to luminosities. Comparison of true and apparent brightnesses lead to distances.) Ultra-great distances are also being measured through observation of Type Ia supernovae (produced when the overflow of mass from a normal star onto a white dwarf companion causes the white dwarf to exceed its support limit and explode). They have closely similar peak brightnesses and, like Cepheids, make fine 'standard candles.'

Ripples in the Cosmic Microwave Background (CMB) combined with Big Bang theory yield all the cosmological parameters. Deep surveys with the Hubble and various ground-based telescopes give the distribution of galaxies and their clusters that devolve from the ripples of mass-energy that earlier created the CMB. X-ray telescopes direct us to the hot matter in the Universe. Chemical compositions of primitive stars give information on the density and distribution of matter and thus on the curvature of the Universe, as do observations of the motions of galaxies. Maximum stellar ages, derived from the theory of stellar structure and evolution, yield lower limits to the age of the Universe.

All the data lead to the same conclusion, that the Big Bang indeed happened and that it is the only viable theory for the origin and evolution of the Universe. The observations also give (with some inevitable argument) a consistent picture of the prime cosmological parameters. The expansion rate of the Universe (the Hubble Constant, H_0) is 73 kilometers per second per megaparsec. (For every million parsecs, 3.26 million light years, the expansion speed relative to Earth increases by 73 km/s.) Twenty years ago, this number was not known to within a factor of two. The error bar is now under six percent. Tip H_0 upside down and you get the age

have struggled to refine the expansion rate so as to date the Universe's birth, trying to fit all observations into a coherent picture. They have indeed done so, learning more and more about the details, while oddly seeming to know less and less of how it all works.

The expansion of the Universe has long been explained by what has become the standard theory, that of the Big Bang. It postulates a specific origin created by a sudden expansion from a highly compacted, hot dense state. The birthdate depends on the expansion rate and the curvature of Einstein's four-dimensional spacetime. The theory is supported by a broad range of observations, the most dramatic of which is the Cosmic Microwave Background (CMB) radiation, the all-pervasive remnant of the original Big Bang 'fireball' now cooled to 2.7 kelvin.

The Numbers

Our knowledge of the overall picture comes from a variety of sources. The Hubble Space Telescope can discriminate Cepheid variables in galaxies

Above: *The Andromeda Galaxy, M31, the closest large spiral galaxy some two million light years away, is easily resolved into bright stars and reddish gaseous nebulae. Image courtesy T. A. Rector and B. A. Wolpa/NOAO/AURA/NSF.*

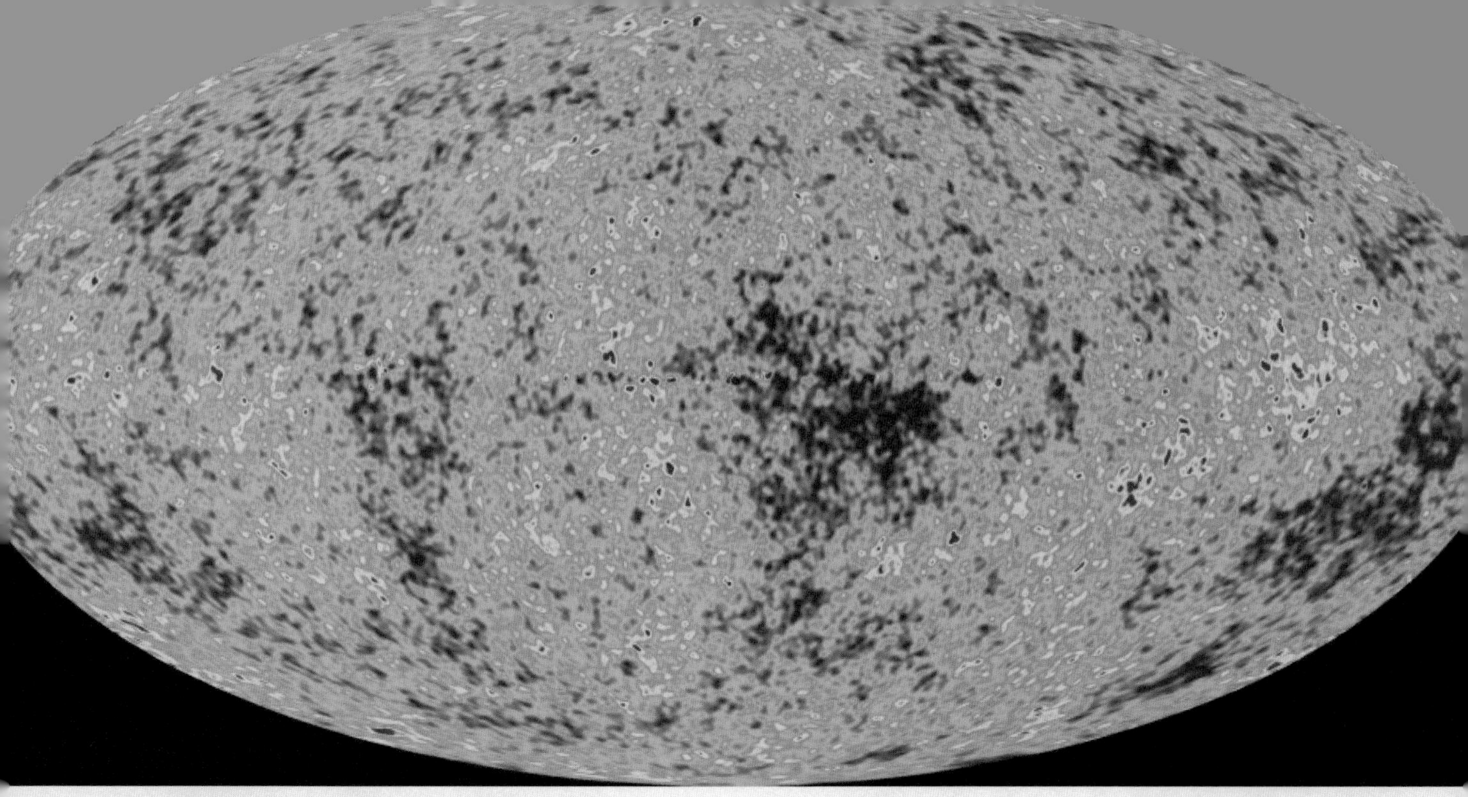

of the Universe since the Big Bang (t_0), 13.5 to 14 billion years, which fits beautifully with the ages of the oldest Galactic globular clusters (12.5 or so billion years) and with current measures from the radioactive decays of uranium and thorium.

Shape

The age of the Universe derived from H_0 requires that we know the 'shape' of the Universe, in the most simple terms whether it is curved outward (in Einstein's spacetime) and thus 'open' and forever expanding; curved inward and 'closed', thus apparently bound to stop expanding and someday collapse; or 'flat', such that Euclid's geometry works (the Universe coming to a halt only after infinite time has elapsed). The search for shape is epitomized by the last letter of the Greek alphabet, Omega, which is the ratio of the current average density of the mass in the Universe compared with the density needed just to close it. Omega is 1 if the Universe is flat, greater than 1 if closed, less than 1 if open. The concordance of data point firmly to 'flat', and therefore to 'inflation', the idea that shortly after the Big Bang the Universe underwent a short burst of amazingly fast expansion such as to enforce flatness. To make Omega = 1 requires sufficient density of mass, and therein lie two of the greatest mysteries of all astronomy. One has been around for decades, while the other is newly visited upon us.

Dark Matter and Energy

Add all the familiar stuff together, stars and the star-birthing interstellar matter. From statistical arguments we can 'count' stars in our Galaxy, and can derive the counts for other galaxies through their luminosities. Stars and related normal 'baryonic' matter (that made of ordinary protons and neutrons) add up to an Omega of a mere 0.6 percent, far below that required for flatness. The Big Bang provided us with hydrogen, helium, and a bit of lithium (with everything else made in stars through nuclear fusion reactions). The amounts are sensitive to the density of baryonic matter, and lead to a much higher value of Omega (as do temperature ripples in the all-sky map of the CMB) of about 0.04. Most baryonic matter has been found to be in gas that pervades clusters of galaxies, the bulk of it primordial, but with a contribution blown out of the member galaxies by supernova explosions. Heated to near-invisibility, into the millions of kelvins, it is known through its X-ray emissions. So far we are short by 96 percent. Let's look for some clues about the rest of the matter in the universe: the unseen part.

Stars orbit within galaxies under the influence of the mass interior to the orbital paths. Stars at great distances from the galactic centers move much too fast for the mass that can be seen. Moreover, galaxies within clusters are moving too quickly as compared with calculations based on visible mass, observations that go back

■ ***Above:*** *An all-sky view of the three-degree Cosmic Background Radiation taken with the Wilkinson Microwave Anisotropy Probe reveals tiny temperature variations, from which cosmologists can find all the numbers that characterize the expanding Universe. From these ancient ripples grew the clusters of galaxies we see today. Image courtesy NASA/WMAP Science Team.*

Dark matter in a huge cluster of galaxies two billion light years away so distorts the fabric of spacetime that it gravitationally lenses even more distant galaxies into weird shapes. Image courtesy ESA/NASA, J. P. Kneib (Caltech/ Observatoire Midi Pyrénées) and R. Ellis (Caltech).

to the 1930s. The stuff that provides the extra gravitational field cannot be seen, cannot be detected in any other way. No one knows what it is, except for the negative, that it is not baryonic. Its presence is clearly indicated by 'gravitational lensing' of the light of distant sources as it passes through intervening galaxy clusters (lensing caused by the distortion of spacetime produced by mass), such data showing that dark matter at least tracks the baryonic variety.

To provide the observed gravitational fields there must on the average be five times as much as there is baryonic matter. By itself, dark matter gives Omega = 0.21. Adding in baryonic matter raises Omega to of about 0.25: far short of enforcing flatness.

For most all the twentieth century, astronomers searched for the cosmic 'deceleration parameter', which told at what rate the expansion is slowing down under constant drag of gravity. Surprise. Not only is it not slowing down, the expansion is speeding up! Distances from Type Ia supernovae clearly show the Universe to be accelerating. Acceleration, however, requires energy, energy that has no obvious origin. Before the Universe was known to be expanding, Einstein postulated an outward 'cosmological force' that would counter the collapse of the Universe through gravity, thus keeping it stable. Further theory plus Hubble's discovery of the expansion voided the concept. It's back, but now as some kind of 'dark energy' whose nature is entirely unknown. Energy and mass (Einstein again) are equivalent through $E = mc^2$. The mass equivalent adds an Omega of 0.75. Add to the 0.25 of baryonic plus dark matter and the total is ONE. The CMB by itself yields the same numbers.

Everything fits, data from all sources fully consistent within small error limits. We live in a flat, accelerating Universe. Observation shows that up to five billion years ago, seen when looking five billion light years away and into the past (with a redshift approaching 1), that the Universe was indeed decelerating, the mysterious acceleration then taking over. Given a positive cosmological force, the Universe is then bound to expand forever, never coasting to a halt (nor would it necessarily even if closed).

Galaxies

Within this framework, the CMB is the oldest and most distant thing we can see, created when the Universe first became transparent to radiation roughly 300,000 years after the Big Bang at a redshift of about 1000. The cooling gas condensed into blobs that began to form stars, hence galaxies, and very quickly too, within a few hundred million years. More than 100 billion of them, each with a hundred billion stars of more, could potentially be seen with current observing capability. Massive black holes at the centers of these systems accreting surrounding gas became the quasars that we see off in the distance today. Our observing capacity takes us back to redshifts of 6, 7, even 10, to less than a billion years after the creation event.

■ *A jet shoots violently from a central two billion-solar-mass black hole in the elliptical galaxy M 87, reminiscent of more distant and ancient quasars. Image courtesy NASA and the Hubble Heritage Team (STScI/AURA)*

Over the aeons, galaxies grew large through successive mergers, one of them becoming our own Milky Way, some of the central black holes reaching well beyond a billion solar masses. Even our Galaxy has a 'small' one rated at a bit over three million solar masses. Theory shows how the ripples in the Big Bang CMB develop into the filamentary strings, walls, and clusters of galaxies that have been observed through extensive large surveys (which are providing so much data that astronomers have founded a 'National Virtual Observatory' that requires no visits to mountaintops). All seems understood. Until we realize once again that 96 percent of the mass-energy of the Universe is 'missing' from our font of knowledge. We have no clue as to what it is. Or for that matter where the Big Bang came from in the first place.

■ *Twin jets 'bipolar flows' emerge from a dark disk that surrounds a buried newly-formed star. Accumulation of dust in the disk may congeal to form planets: a "solar system" waiting to be born. Image courtesy STScI and NASA.*

Stars

We at least do know what the stars are and that they shine, and are supported, by internal nuclear fusion, which creates energy from the fusion of light chemical elements into heavier ones (starting with hydrogen to helium). We know that they are born from condensations in the interstellar medium, and have a pretty good idea how the process works. Massive-star winds and exploding stars - supernovae - compress the thin, lumpy, dusty interstellar gas to the point where some of the randomly created overdense blobs can contract under the force of their own gravity. As the contraction causes a new star to spin faster, some of the infalling mass gets spun out into a dusty equatorial disk that directs mass to flow back outwards along the rotation axis in a bipolar flow that announces 'a new star lives here'.

New observations of massive star clusters show that the most massive stars that can be formed hover between 120 and 200 solar masses (the best estimate closer to the lower number). Such great lights shine with the power of millions of Suns. The lower mass limit is more mysterious.

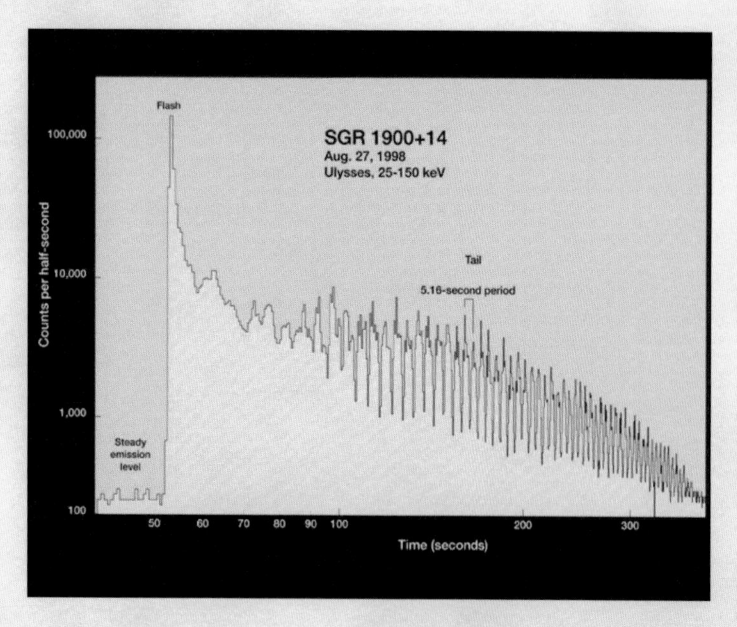

Down to about 8 percent of a solar mass, stars can run the chain of nuclear reactions from hydrogen to helium (such stars 30,000 times less luminous than the Sun), even if just to helium-3, but below that limit they cannot become hot enough inside. These 'substars', called 'brown dwarfs', are being found in profusion by infrared survey telescopes. We do not know how low they can go. They may even overlap the masses of the giant planets. There are so many dim, cool stars (down to below 1000 kelvins) that two new spectral classes, L and T, had to be added to the classic OBAFGKM sequence.

High Mass

When fusion brings their cores to iron, high mass stars, those above 8 or 10 solar, blow up as 'Type II' supernovae, and the cycle of creation starts all over again. Such explosions (as well as those of Type Ia) create all the natural chemical elements, which then find their way into the next stellar generations. Stars of up to 25 to 40 solar masses collapse into neutron stars that compress the mass of a Sun into the size of a small town. Higher mass stars may collapse into stellar-size black holes from which even light cannot escape. Neutron stars (most of which are quiet and hard to observe) make the myriad rapidly rotating magnetic (a trillion times Earth) pulsars that are known to flock the Galaxy.

They also make newly-recognized 'magnetars' that have magnetic fields that are 100 to 1000 times those of normal pulsars. Rotation twists the internal fields so much that they break through the magnetar's crusted surfaces (so goes theory). Field re-arrangement then produces a massive second-long burst of gamma rays that is followed by a slow decay. The flares from these 'soft gamma ray repeaters' ('SGRs') are so powerful that even at distances of tens of thousands of light years, the radiation overwhelms Earth-orbiting satellites and ionizes the Earth's upper atmosphere.

SGRs are not the source (or at least the main source) of the 'gamma ray bursts' ('GRBs') that come to us daily from the far reaches of the Universe. They seem instead to have two quite different origins. Longer bursts, those that last more than about two seconds, appear to be caused by super-energetic 'hypernovae' that may make stellar black holes (or perhaps magnetars). Radiation beamed out the rotation poles at near the speed of light could, if the exploding star is close enough, cause massive damage to Earth. Short GRBs are more likely caused by mergers of binary neutron stars, binary black holes or a black hole and neutron star - and even a few by SGRs.

Lower Mass

More ordinary stars, those like the Sun, die as common white dwarfs, which are old, dead nuclear-burning cores about the size of Earth that could get no farther in fusion than carbon and oxygen (a tiny, more massive fraction going to neon and oxygen). In the preceding giant stages, convection coupled with winds can bring up freshly made chemical elements from the nuclear furnaces that add to those coming from supernovae. The result is that the young stellar generations of the Galactic disk ('Population I') have more heavy elements in them than do the older generations of the halo ('Population II'), which surrounds the disk. The ancient globular clusters have metal abundances only a hundredth that of the Sun, while the record is an iron-to-hydrogen ratio a mere quarter-millionth solar. Theoretical calculations of chemical formation and return to interstellar space are good enough to come close to modeling the solar composition.

Since the Big Bang created only the three lightest elements, there should be a zero-metal 'Population III.' That Pop III is not seen suggests that the earliest stars were massive and quickly blew up to salt the interstellar gases with just enough metals to account for those seen in the oldest stars. There are simply none of the most primitive stars left to find.

Above: *Soft gamma ray repeater SGR 1900+14 let out a gamma ray flare so mighty that it ionized the Earth's upper atmosphere. As the flare died away, observers saw pulses from the rotating magnetar, whose powerful magnetic field had rearranged itself. Image courtesy Kevin Hurley, Univ. of Cal. at Berkeley, and NASA/Marshall.*

Planets

Planets have long been thought to condense from the disks surrounding forming stars, first through dust accumulation and then (in the case of giant planets) through gas accretion. The theory explains both the form of the Solar System and the chemical compositions of its planets. Compared with a solar lifetime, formation is quick, taking only the order of 100 million years. Debris disks surrounding other stars like Vega, Fomalhaut and Beta Pictoris support the concept.

The most exciting arena, however, is in the discovery not so much of the disks, but of real planets orbiting other stars, the field as explosive as the supernovae that provide the heavy elements of which the planets (or their cores) are made. Astronomers have now found around 200 'exoplanets' that include 20 multiple-planet systems (with a few brown dwarfs on the side). The vast majority have been revealed through sensitive radial velocity observations, wherein we observe slight back-and-forth motions of a star as its planet orbits. A significant number of these are 'hot Jupiters', giant planets formed at larger distances from the stars and subsequently brought in by friction with the remnants of the disks. Surely, some planets have even spiralled into their parent stars. Longer time baselines are now yielding systems and planets a bit more like our own. The low-mass record for the technique is just nine Earth masses for a planet that orbits a star called Ross 780 (a low-mass class M dwarf). Stay tuned, however, new results are announced on a regular basis and this record will be broken.

Other techniques are rapidly developing or promising, including direct imaging, observations of planets that transit in front of their parent stars (thus causing dips in stellar brightness), detection of side-to-side (astrometric) motions, and 'microlensing', gravitational lensing of distant stars by small bodies. The latter method captured the low-mass record with a planet of but 5.5 Earth masses. While not quite direct imaging, infrared radiation has been detected from two orbiting hot Jupiters.

Someday we will recover real 'earths.' Someday we may even know something of what is on them (though the Search for Extraterrestrial Intelligence - SETI - has so far come up blank). Someday way may even know what dark matter and dark energy are. So even though we sometimes seem to know less and less as we learn more and more, we still strive to understand our surroundings, whether in near space or in far, to find our place in the expanding Universe and our personal relationship with the Big Bang that started it all.

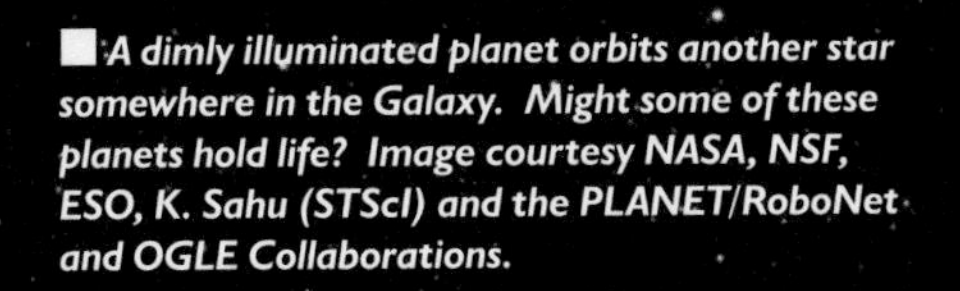

A dimly illuminated planet orbits another star somewhere in the Galaxy. Might some of these planets hold life? Image courtesy NASA, NSF, ESO, K. Sahu (STScI) and the PLANET/RoboNet and OGLE Collaborations.

4
First Views of the BIRTH of a SUPERNOVA Remnant...

Bunny often wondered what the Human Astronomers found so interesting on their screens...

As we approach the 20th anniversary of the discovery of the famous 1987 supernova in the Large Magellanic Cloud, world-renowned expert *Richard McCray* gives us the most recent update regarding the search for the central compact object, and the dramatic flaring of the gaseous ring surrounding the supernova.

...SUPERNOVA 1987A

SUPERNOVA 1987A is the brightest supernova since Kepler's supernova of 1604 AD, the first to be observed at all bands of the electromagnetic spectrum, from radio to gamma rays, and the first to be observed through its flash of neutrinos. Today, as we approach the 20th anniversary of its discovery (23 February 1987), SN1987A is giving us our first view of the birth of a supernova remnant, the glowing distribution of gas that results from the impact of the supernova debris with its circumstellar environment.

SN1987A was first observed through its optical display, which resulted from the radioactive decays of newly synthesized isotopes such as 56Ni (Nickel), 56Co (Cobalt), 57Co, and 44Ti (Titanium). The luminosity in positrons and gamma rays from these decays was absorbed in the inner debris and converted to optical radiation. The optical luminosity of the supernova reached a maximum value of about 250 million Suns about 3 months after the supernova was first detected, just about the time when the debris was becoming transparent. X-rays and gamma rays were also detected from SN1987A several months later, when the debris began to become transparent to these wavelengths as well. The integrated energy of the optical light from SN1987A is about 10^{49} ergs (an erg is a unit of energy and equivalent to 10^{-7} Joules), about what would be expected from the radioactive decay of 0.07 solar masses of newly synthesized 56Ni.

Today, the optical light from the inner debris has diminished by 100 million times. It is still visible in images from the Hubble Space Telescope. Its irregular shape could be due to the distribution of radioactive elements and/or the distribution of

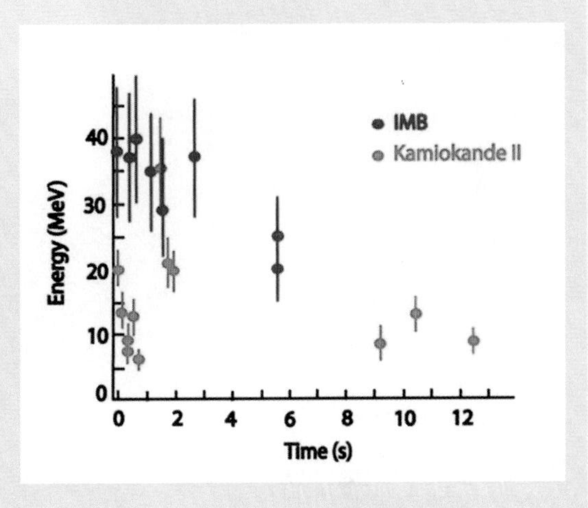

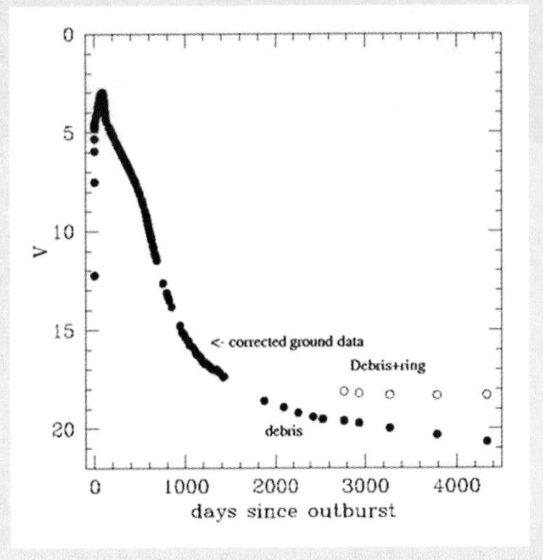

Top: *The neutrino pulse from SN1987A as observed by detectors in Ohio (IMB) and Japan (Kamiokande II). The typical energies of the neutrinos differ in the two experiments because the detectors have different sensitivity thresholds. Illustration courtesy IMB and Kamiokande consortium.*

Above: *The optical light curve of the SN1987A. Illustration courtesy D.McCray.*

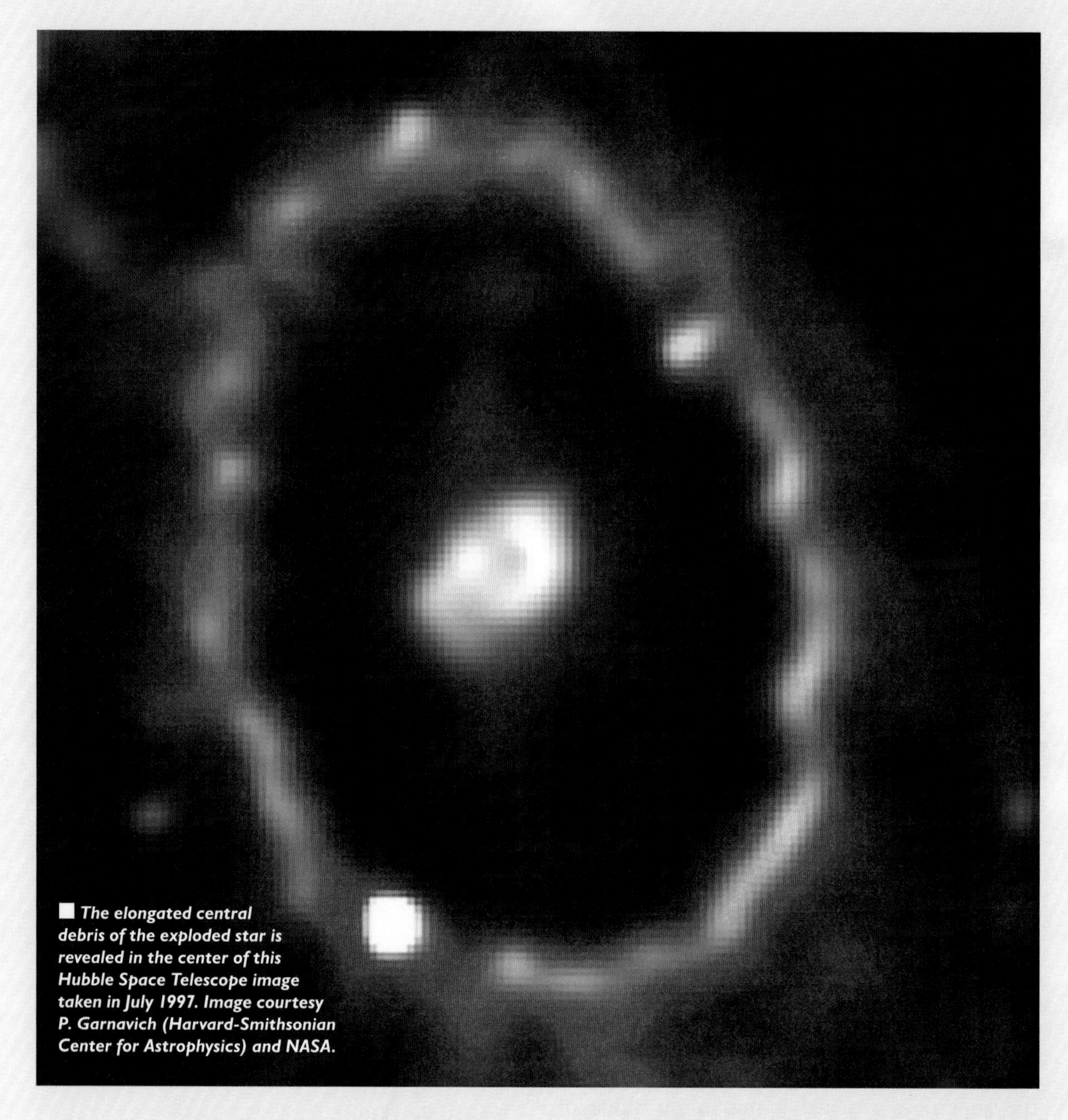

The elongated central debris of the exploded star is revealed in the center of this Hubble Space Telescope image taken in July 1997. Image courtesy P. Garnavich (Harvard-Smithsonian Center for Astrophysics) and NASA.

absorbing dust clouds in the inner debris. This shape is expanding with radial velocity of about 3,000 km/s. Spectra of this source indicate that the emitting gas is very cold, less than 300 Kelvin, making it perhaps the coldest optically emitting object known to astronomers.

Searching for the Central Object

Soon after SN1987A was discovered, physicists examined records from deep underground detectors in both Japan and the US and found evidence of a flash of neutrinos from SN1987A. (See Figure 1) The flash had the total energy of about 3×10^{53} ergs (the Sun emits 3.8×10^{33} ergs per second by comparison), and lasted about 4 seconds, reaching a temperature of $kT \sim 4$ MeV. This is what was expected to result from the collapse of the core of a star to form a neutron star.

(Note: It is customary in plasma physics to express temperature in electron volts (here, million-electron volts, or MeV), which is derived from k, the Boltzmann constant, and T, the temperature in Kelvin. The Boltzmann constant is number that relates temperature to energy. 1 keV = 11.6 million Kelvin)

Although we are certain that such a core collapse occurred, we have no further evidence to this date of the compact object that we expect to reside at the center of the debris. The upper limit to the integrated luminosity of a compact source at the center of SN1987A is 100 Suns. This result is a mystery. By comparison, the neutron star in the Crab Nebula still has luminosity of

100,000 Suns, some 950 years after it was formed. Why should a newborn neutron star be so faint? In fact, the upper limit to the observed luminosity is only slightly greater than the lower limit that would be expected theoretically from a cooling neutron star that had no spin and no magnetic field.

One possible explanation for the absence of any evidence for a neutron star is that, shortly after its formation, sufficient material fell back onto the neutron star to cause a further collapse to form a black hole. If so, the accretion of supernova debris must have subsequently diminished to a very low value (less than one billionth of a solar mass per year); otherwise we would see evidence of radiation from the accreting gas.

The supernova debris is still opaque in the radio and soft X-ray bands. It may also be opaque at optical and infrared wavelengths as well if a foreground cloud of dust in the debris covers the center. We may have to wait decades for the supernova debris to thin out enough to see the very faint compact object that must be there.

The Circumstellar Rings.

The image below shows the remarkable system of three circumstellar rings of gas around the supernova. These rings are actually nearly circular but appear elliptical because the entire system is inclined to the south by approximately 45 degrees. The inner ring has a radius of approximately 0.8 light years and is coplanar with the supernova. The outer loops are about three times as large. They are coaxial with the inner ring but not coplanar. The inner ring is expanding at about 10 km/s, while the outer loops are expanding roughly three times faster. Dividing the radial distance of the rings from the central source by the radial velocity gives a time of approximately 20,000 years, since the rings were ejected from the progenitor system.

Element abundances determined from the spectrum of the ring are typical of the interstellar gas of the Large Magellanic Cloud, except that the ratio of nitrogen to oxygen is elevated by eight times. The elevated nitrogen abundance can be explained if the gas in the ring has been

■ *FSN1987A and its system of three circumstellar rings, as observed by the Hubble Space Telescope in 1992. The spot in the center is the inner debris of the supernova itself, which is glowing by virtue of the radioactive elements formed there during the explosion. The three rings are actually nearly circular but appear elliptical because the entire system is inclined by approximately 45 degrees. The inner ring is equatorial. The two outer loops are approximately coaxial with the inner ring but not coplanar. Image courtesy P. Challis (CfA).*

expelled by the progenitor star. In a massive star such as the progenitor of SN1987A, carbon and oxygen are converted into nitrogen through the CNO cycle of hydrogen burning in the stellar core and then, through convection, is raised to the stellar atmosphere.

The origin of the circumstellar rings remains a mystery. We see similar bipolar geometries in many planetary nebulae, for which the standard explanation is that the nebulae represent mass loss from a binary star progenitor. Perhaps the progenitor system of SN1987A was a binary system, and the rings were ejected during a merger event that took place some 20,000 years ago.

Such an event provides a natural scenario for forming the inner ring. During the merger, a fraction of the gas flowing from the more massive progenitor to the less massive star would naturally be expelled in a spiral outflow that would wind into a disk. Afterwards, the ionizing radiation and stellar wind from the progenitor would erode a large hole in the disk. Then, the flash of ultraviolet and soft X-rays from the supernova would ionize the inner rim of this hole and cause it to light up. Indeed, we know that the total amount of ionized gas in the inner ring (about 0.07 solar masses) is just what we would expect from such a flash. The ionized gas in the rings has continued to glow up to the present time, fading slowly because its density (a few x 10^3 atoms cm-3) is sufficiently low that the recombination and cooling timescales span several years. Likewise, the outer loops are glowing because they were ionized by the supernova flash. But we have no good hydrodynamical model to account for their morphology.

We do know that the glowing triple ring system of SN1987A is just the tip of the iceberg. Rapidly evolving light echoes near the supernova (but beyond the inner ring) were observed during the first few years after outburst. These echoes result from the scattering of the supernova light by dust grains. Assuming a typical gas-to-dust ratio, the brightness of these echoes indicate that several solar masses of circumstellar gas were probably also ejected by the supernova progenitor system. No doubt, the distribution of this gas contains the clues we need to understand the ejection event. Unfortunately, the echo data were not good enough to map the distribution of this gas completely. This gas was not ionized by the supernova and so is invisible now. Fortunately, as we describe below, events taking place during the next several years will give us another chance.

The Newborn Supernova Remnant

The expanding debris of SN1987A has kinetic energy of about 1.5 x 10^{51} ergs, roughly two orders of magnitude less than the energy released as neutrinos but two orders of magnitude greater than that of the optical display. To put this massive amount of energy into context,

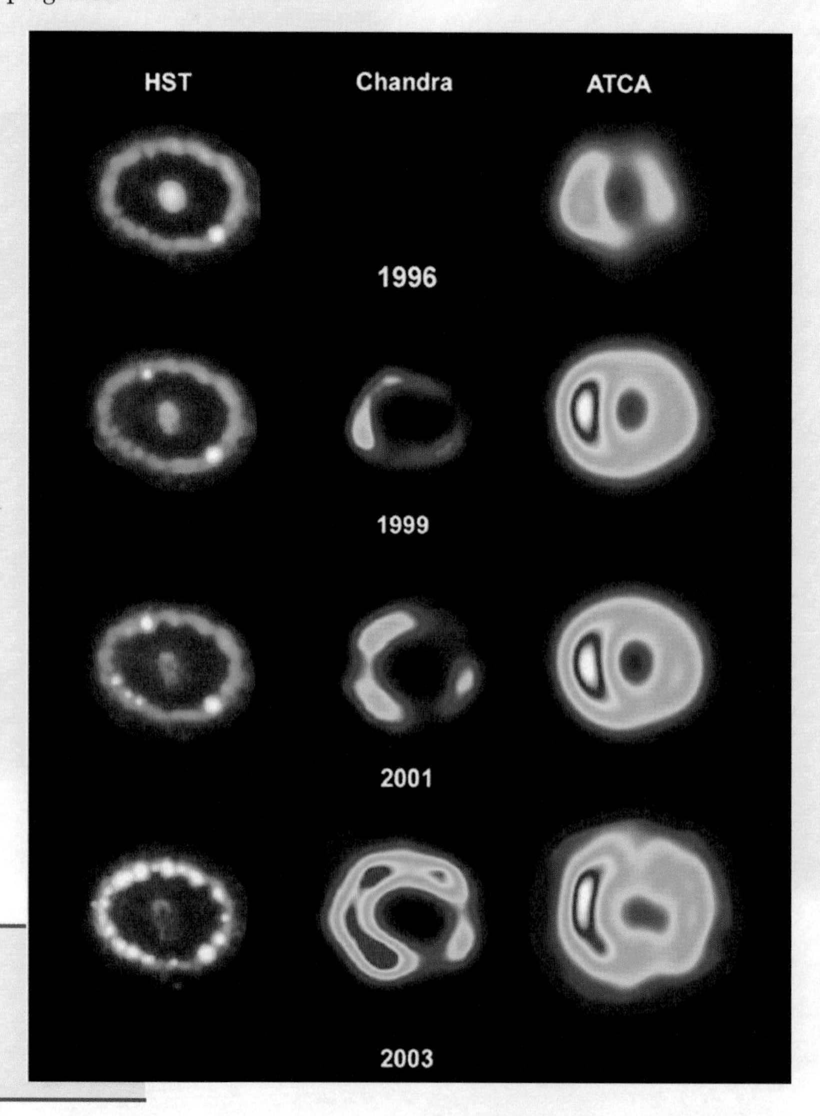

Right: *Montage illustrating the evolution of the images of SN1987A in the optical (left column), X-ray from Chandra (center) and radio from the ATCA telescope (right). Images courtesy P. Challis (for HST), Sangwook Park (for Chandra) and D. Manchester (for ATCA).*

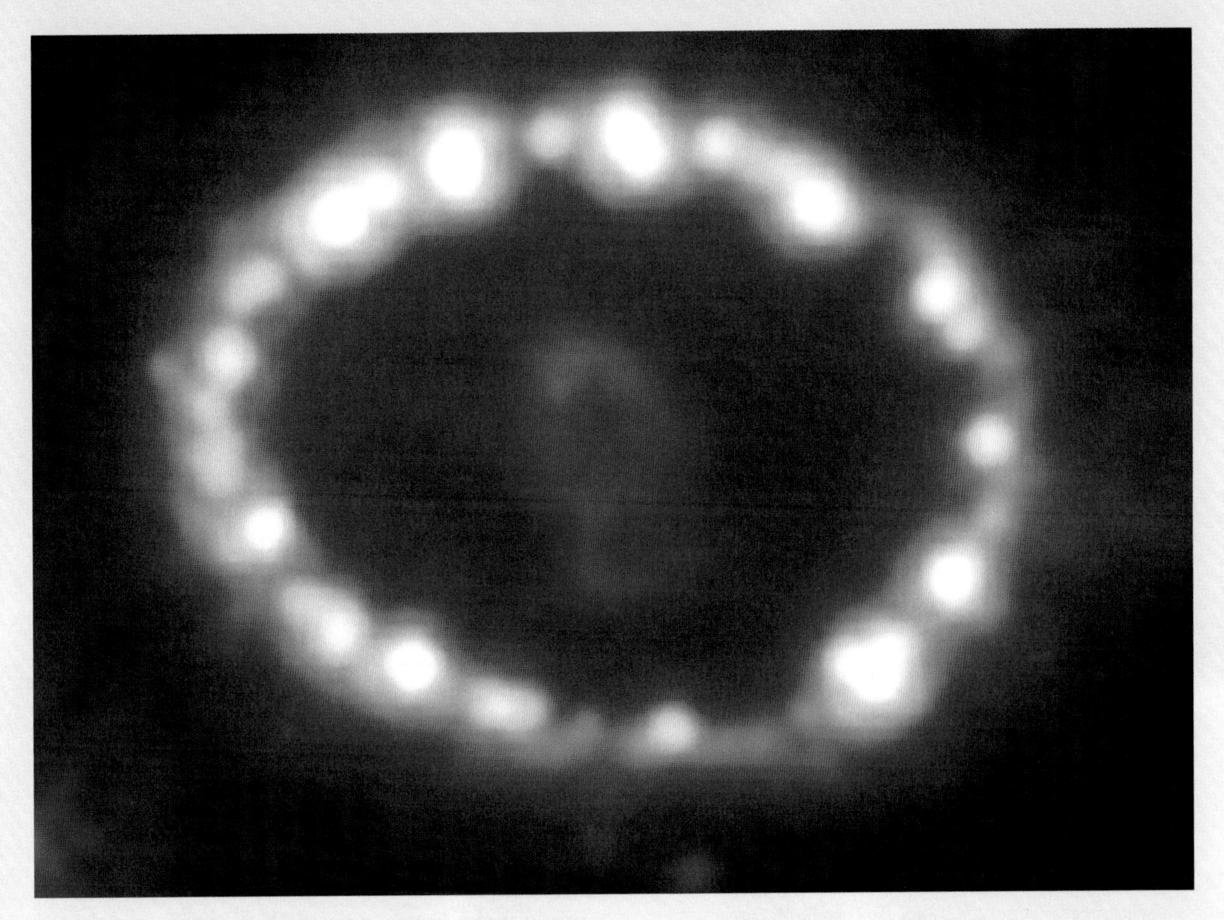

10^{51} ergs is equivalent to a billion megatons of TNT going off every second for a trillion years. Most of this kinetic energy is contained in the invisible expanding outer envelope of the supernova. It will be converted to radiation only when the supernova debris (approximately 10 solar masses) strikes a comparable mass of circumstellar matter. The shocks resulting from this impact will raise the temperatures of both the circumstellar gas and the supernova debris to tens of millions of degrees, and this gas will radiate its energy mostly as X-rays.

This process is taking place now. About 1,200 days after the supernova explosion, scientists detected a steadily brightening source of X-rays and non-thermal radio emission. The radio image showed that the source was an annulus located inside the inner circumstellar ring. The annulus was expanding with radial velocity of about 3,500 kilometers per second. Evidently, the radio emission was synchrotron radiation from relativistic electrons accelerated at shocks formed where the supernova blast wave was impacting relatively low density circumstellar gas inside the inner ring (about 100 atoms per cubic centimeter).

The impact became much more dramatic when the blast wave began to strike the inner ring. The first evidence of this encounter was the 1995 appearance of the rapidly brightening "Spot 1" on optical images of the inner ring taken with the Hubble Space Telescope.

Spectra of Spot 1 showed that the optical emission was coming from gas in the inner ring that was being heated and crushed by shocks transmitted by the blast wave into the denser gas of the inner ring (about 10,000 atoms per cubic centimeter). The widths of the spectra lines, broadened by Doppler shifts produced by the speed of the gas, indicated that the shocked gas responsible for the optical emission had velocities of about 150 kilometers per second, much less than that of the blast wave that was overtaking the inner ring.

The fact that the initial optical brightening was confined to a single spot, unresolved in the Hubble Space Telescope but located slightly inside the main body of the ring, indicated that Spot 1 was a thin finger of dense gas protruding radially inward from the ring. A few years later, several more such hotspots appeared, and today we see more than 20 hotspots encircling the entire ring. As the blast wave overtakes the ring, we expect that the hotspots should merge into a single continuous ring, but that hasn't happened yet.

Above: *The brilliant chain of bright hot spots encircling the elongated central debris of the exploded star is revealed in the center of this Hubble Space Telescope image taken on 28 November 2003. Image courtesy NASA, P. Challis, R. Kirshner (Harvard-Smithsonian Center for Astrophysics) and B. Sugerman (STScI).*

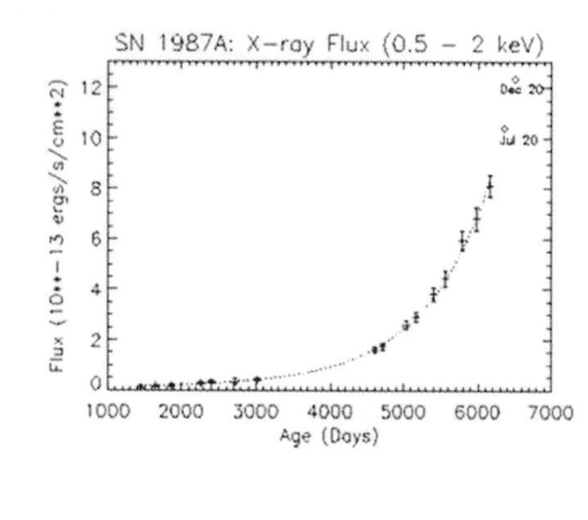

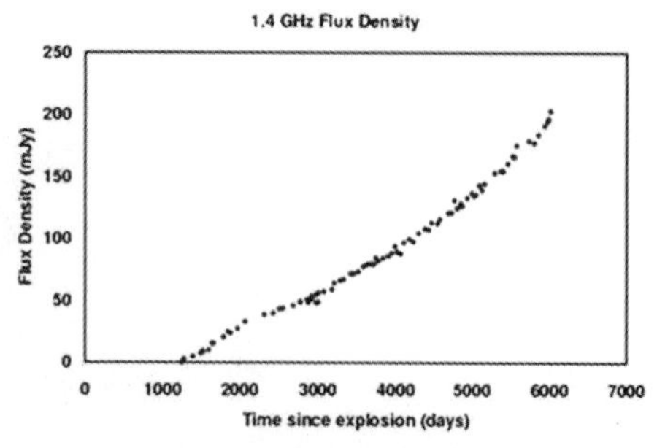

Light curves of X-ray, radio, and optical emission from the impact of SN1987A with its circumstellar matter. Illustrations courtesy Sangwook Park and D. McCray.

The protrusions of dense gas responsible for the hotspots were certainly present before the supernova event. Perhaps they were formed as the stellar wind of the progenitor star eroded a hole in a circumstellar disk that may have been expelled during a merger event. But we haven't yet identified the mechanism by which these protrusions were shaped. The hotspots have remarkably regular spacing, and this suggests that they were formed by some sort of instability with a characteristic wavelength comparable to the thickness of the ring.

Soon after the optical hotspots appeared, the soft X-ray emission from SN1987A began to brighten at an accelerating rate. Today, the X-ray source is doubling in brightness every one and a half years or so. X-ray images of SN1987A taken with the Chandra telescope show an annular source that is nearly coincident with the inner ring, with the brightest arcs of X-ray emission coinciding roughly with the brighter hotspots. The X-ray spectrum is dominated by emission lines of hydrogen-like and helium-like ions of Nitrogen, Oxygen, Neon, Magnesium, Silicon, and certain features produced by Iron ions. The X-ray emitting gas has temperatures in the range $kT \sim 0.5 - 3$ keV.

The broadened line profiles indicate that the gas has velocities in the ranging from 340 to 1,700 kilometers per second, considerably faster than the gas responsible for the optical emission from the hotspots. Evidently, when the blast wave strikes the hotspots, shocks transmitted into the protrusions have a very wide range of velocities depending on the density of gas in the protrusions and the angle of incidence of the blast wave. We see optical emission only from the slower shocks because these are the only shocks for which the shocked gas is dense and cool enough to emit substantial optical radiation. Faster shocks will heat the gas to temperatures sufficiently high to emit X-rays but relatively little optical radiation.

The Future

During the next decade, we expect the newborn remnant of SN1987A to become at least 10 times brighter than it is today, as the blast wave overtakes the entire inner circumstellar ring. This brightening is likely to cast some light on one of its outstanding mysteries: what shaped the circumstellar matter? To address this question, we need to see the distribution of all the gas expelled by the supernova progenitor, not just the inner ionized skin of this gas that is represented by the three rings we see today. The rapidly brightening X-rays from the impact will make this possible. They will ionize invisible matter beyond the circumstellar rings and cause it to glow, just as the X-rays from the initial supernova flash ionized the rings. When this occurs, we will be able to map this matter and its kinematics in more detail than was possible by observing the initial light echoes.

Within a decade, the Atacama Large Millimeter Array (ALMA) will provide us with images of the nonthermal radio emission from SN1987A having angular resolution (less than 0.05 arcseconds) far better than the radio images (about 0.5 arcseconds) that we are able to obtain with the Australia Telescope Compact Array (ATCA). With such images, we will be able for the first time to identify and track the motions of the shocks where the relativistic electrons emitting this radiation are accelerated.

Where is the Pulsar?

We don't know when we will be able to detect the compact object at the center of SN1987A. Perhaps the best hope is to see such a source at infrared or millimeter wavelengths where the supernova debris is likely to become transparent first. It will become increasingly difficult to detect a faint central X-ray source embedded within the rapidly brightening X-rays from the impact. If the central

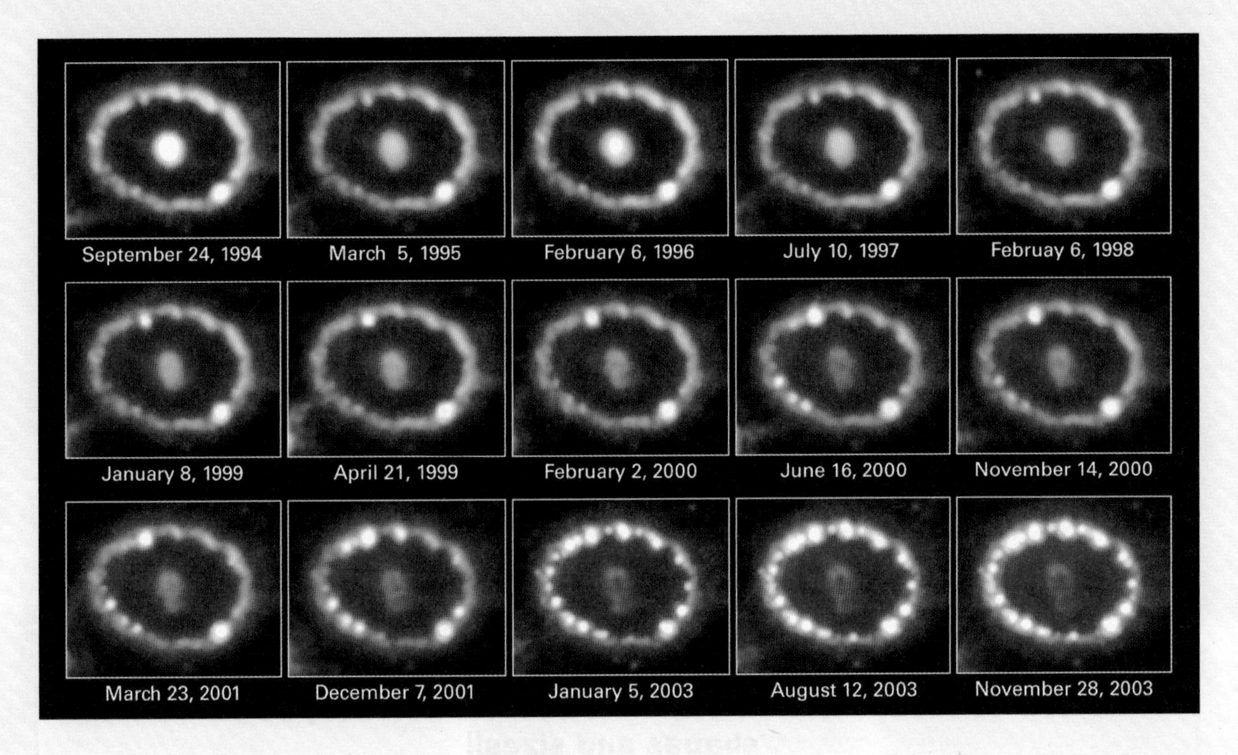

object is a rotating neutron star, one might use time series analysis to detect a faint pulsing source against a brighter background.

One of the main goals of supernova research is to measure the distribution of newly formed heavy elements in the supernova debris. This will become possible in SN1987A when these elements begin to cross the reverse shock. However, it will probably take a few more decades before this event commences.

The diagram below illustrates the hydrodynamics that we believe are now taking place in the remnant of SN1987A. The blue represents the unshocked supernova debris. The outer part of this debris consists mainly of cold (less than 100 Kelvin) neutral hydrogen atoms.

The hydrogen atoms cross a reverse shock, where they are decelerated from about 10,000 kilometers per second to 3,000 kilometers per second and become ionized and heated. The shocked supernova debris drives a blast wave ahead into circumstellar matter, and eventually strikes the denser gas in the inner circumstellar ring.

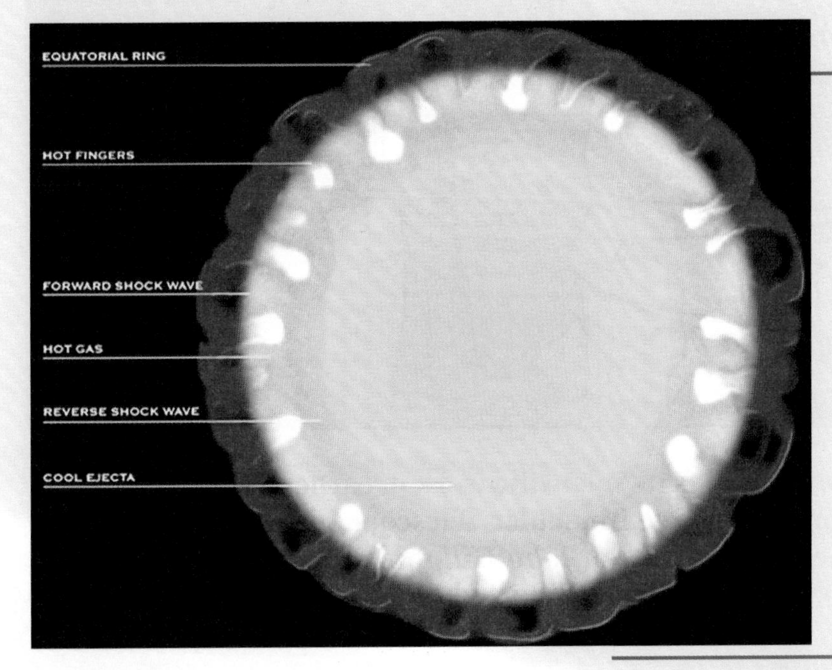

Top: *The evolution of the hotspots around the expanding supernova as recorded by the Hubble Space Telescope from 1994-2003. Images courtesy NASA, P. Challis, R. Kirshner (Harvard-Smithsonian Center for Astrophysics) and B. Sugerman (STScI).*

Left: *Diagram illustrating the impact of the debris of SN1987A with its circumstellar matter. The freely expanding supernova debris (blue) is suddenly slowed from a velocity ~ 10,000 km/s to ~3000 km/s at the reverse shock surface. The shocked supernova debris and circumstellar matter (yellow) drive a blast wave ahead into the circumstellar matter. The blast wave strikes protrusions (white) of relatively dense matter extending inward from the inner circumstellar ring. Illustration courtesy NASA/CXC/M.Weiss.*

5
Unveiling a NEW GENERATION of Extremely Large TELESCOPES...

The new telescope had arrived... It seemed telescopes came in all shapes and sizes!!

As a giant new telescope begins its scientific work in earnest, the Director of the newly completed Large Binocular Telescope, *Richard Green*, and his colleague, Technical Director *John Hill*, provide an inside look at this revolutionary telescope and observatory.

...The Large BINOCULAR TELESCOPE Achieves First Light

THE LARGE Binocular Telescope (LBT) stands on the threshold of its scientific career as the first of the next generation of extremely large telescopes. The telescope achieved first light on 12 October 2005, by imaging the nearby edge-on spiral galaxy, NGC 891. The light was gathered by the first of two primary mirrors of 8.4-meters (27.6 feet) diameter and focused on a wide-field CCD mosaic camera at prime focus. This image below marked the very first step toward full realization of the power of this massive instrument.

The goal of the astronomers in the LBT partnership is to attack some of today's most fundamental problems in astrophysics. Astronomers intend to exploit the power of the binocular aspect of the telescope in two ways. One is to use the collecting area of the two optical channels in parallel to feed pairs of light analyzing instruments. There will be a pair of wide-field CCD mosaic cameras ('LBC' – the Large Binocular Cameras), one at each prime focus station. Observations through multiple color filters are the keys to distinguishing the temperatures of

■ *First-light image with LBT, the edge-on spiral galaxy, NGC 891. The image was taken through a blue filter with a total exposure time of 300 seconds.*

stars, the redshifts of galaxies, and the presence of objects with markedly different color-color ratios in different wavebands compared with stars, such as quasars. One camera works best in ultraviolet through green, the other in yellow through far red. Multi-color surveys will therefore take half the time by observing the same field simultaneously in two passbands.

There will be two pairs of wide-field spectrographs, one pair for optical light and one pair for the near-infrared. Both will employ custom-drilled focal plane masks, which provide a pattern of slits exactly matched to the distribution of faint targets (stars or galaxies). When the telescope is perfectly aligned on the field, the instrument pairs will record a total of some 40 faint objects simultaneously. With multi-hour exposures and large telescope time at a premium, this multiplex advantage is critical for studies of groups of faint objects. The immense light-gathering power of LBT will allow spectroscopy of proto-galactic clumps that lie near the edge of the visible universe - so far away their starlight is redshifted into the near-infrared. The theory of galaxy formation by coalescence of smaller units will be directly tested, along with measuring the degree to which galaxies were clustered with each other in large-scale structures at early cosmic times. An additional high-resolution, high-stability spectrograph will be on a climate-controlled optical bench, fed by fiber optics from the telescope. The variations of solar activity that impact life on earth will be set in context by its exquisitely detailed study of the magnetic fields and spot activity on other stars.

The unique and powerful mode of LBT will be in combining the two beams of incoming light in phase. This combination, called interferometry, allows the resolution of the telescope to increase until it is limited by diffraction in the telescope aperture. With the two primary mirrors on a common and stiff mounting structure, their light beams can be combined coherently to provide a genuine field of view. This situation is in contrast to combining the light from separate telescopes; the result in that case is yet higher resolution because of the greater separation, but a tiny field of view in the immediate vicinity of the field center.

With diffraction-limited resolution nearly ten times greater than the Hubble Space Telescope, LBT will image the remnant disks of gas and dust in newly formed solar systems, and detect the gaps plowed out by nascent planets. This will contribute to our understanding of how planetary systems formed and how they evolve. The mid-infrared interferometer will make images and can also combine the beams in a way to null out the light from a bright central star, revealing faint details in its surroundings, such as proto-planetary disks or actual planets. The near-infrared interferometer is an ambitious instrument that will realize the full potential of the LBT for dissecting the nuclei of galaxies and stellar nurseries. The nuclear regions of nearby galaxies will be explored with that resolution, revealing the powerful influence of supermassive black holes on the distribution of stars and gas.

The Primary Mirrors

LBT is the pathfinder for the next generation of extremely large telescopes. Every technical aspect of the project is visionary and state-of-the-art. The giant primary mirrors represent the culmination of developments at the University of Arizona's Steward Observatory Mirror Lab to produce borosilicate honeycomb mirrors. To make sharp images, the front surface of a telescope mirror must have a carefully controlled figure and be held within a fraction of a degree of the ambient air temperature, so as not to create thermal disturbances. The design of the 8.4-meter diameter mirror accomplishes both of those goals. The structure is stiff because the mirror is formed as thin front and back plates supported internally by a network of honeycomb cells of thin vertical glass walls. The front surface is 28 millimeters thick (about an inch), and comes to thermal equilibrium with the air very quickly. It is helped along by forced ventilation of thermally controlled air into the back of each of the honeycomb cells.

The production of the two 8.4-meter primary mirrors is a remarkable technical achievement. Custom-cut refractory cores are bolted down to the floor of the furnace to create the mold to yield the honeycomb pattern and the 'deep dish' (f/1.14) shape of the mirror. The borosilicate glass is type E6 from Ohara glass works in Japan. Mirror Lab team members load 18,600 kg (41,000 pounds) of that glass into the furnace on top of the refractory mold. The entire furnace is spun at 6.8 revolutions per minute to maintain a uniform thickness of the parabolic front surface as the glass melts. The sides of the mold must be restrained by strong metal bands (with a high melting temperature) to resist the centrifugal force of the load of melting glass. At 1,150

degrees Celsius (2,100°F), the glass becomes viscous like honey, and flows around the cores. It is then slowly annealed by gradually reducing the heat in the furnace. After 3 months, the top of the furnace is lifted off, and the resulting glass blank can be inspected.

Handling an 8.4-meter blank is not a trivial activity. The mirror must be hung vertically for a power wash to clean out all the refractory mold material in each honeycomb cell. To get it there, the walls of the furnace are removed, and 36 steel pads are glued to the front surface of the mirror blank. The mirror is lifted with a star-shaped fixture and then attached to a large ring that allows it to be turned on edge. After clean-out, the mirror undergoes its lengthy polishing process. The University of Arizona Mirror Lab devised precision polishing techniques based on computer control of the stresses applied to the polishing tool, or lap. The resulting figure is so smooth that no point across the final surface of the mirror varies from the desired shape by more than 25 nanometers - a millionth of an inch.

That excellent surface figure is maintained in operation by coupling the mirror to a steel cell through a network of 160 pneumatic actuators and a spacing reference system of six adjustable 'hard points'. The actuators compensate for the changing gravity vector of supporting the glass as the telescope moves from zenith to horizon, and for natural bending of the mirror plus cell structure that would distort an image from near-perfect quality.

The mirrors had to be shipped from their place of production under the stadium in Tucson to the Observatory site on the 3,190-meter (10,470-foot) Emerald Peak summit on Mt. Graham in southeastern Arizona. The mirror in its shipping container sat flat on its trailer, taking up two lanes and requiring a uniformed escort for its 200-kilometer (124-mile) trip to the base camp in Safford, Arizona, over interstate and state highways. The container was then lifted by crane onto the special, tipped frame custom-built for transport up the 48 kilometers (30 miles) of steep and treacherous mountain road to the observatory facility at a typical speed of 2 km per hour (1.2 miles per hour). Precision Heavy Haul of Phoenix won the national award for 2004 for the most challenging delivery in the country. On site, the mirror was removed from its container by a vacuum lifting fixture, and

■ ***Top**: The second 8.4-meter primary ascends the mountain grade in its shipping container in September, 2005.*

■ ***Above**: The network of 160 actuators attached to the rear of the 8.4-meter mirror seen as it is removed from its shipping crate.*

■ ***Below**: Moving the 8.4-m primary with the vacuum lifting fixture from its shipping container into its cell in the mountain high bay staging area. The bell jar for aluminizing the mirror is the darker red structure hanging vertically in the upper right.*

placed in its steel cell. The large overhead crane then lifted the mirror and cell from the high bay on the ground floor up to the telescope structure, where the mountain crew bolted the cell in place on the telescope structure. At the time of first light last October, only the 'left-hand' mirror was in place. Just afterwards, the second primary was installed, making the telescope truly binocular. The 'right-hand' mirror was aluminized in January of 2006; 'second light', imaging with both channels and two LBC cameras is anticipated in winter of 2006.

The Telescope Structure

The mount for the Large Binocular Telescope is unique among the modern giant telescopes. All other facilities support a single large primary. The LBT mount was conceived and designed explicitly to support two enormous primary mirrors and auxiliary optics stiffly. The coherent combination of the two telescope beams into one image requires maintaining the difference in path length that the light travels from one side versus the other to be constant to a fraction of a single wave of light, at the level of 1/10 of a micrometer. Disturbances by wind cannot shake the telescope structure and optics mountings beyond the interferometer's ability to compensate by adjusting a miniature 'slide trombone' to match (i.e., phase) the light waves reflected from each side.

■ ***Above**: Both primary mirrors have been installed on the telescope structure and aluminized. The telescope is pointing at the horizon in this view.*

The resulting structure is very stiff, and to get that way, it is massive. The base is a concrete pier 20 meters (66 feet) high and 14 meters (46 feet) in diameter, resting on the mountain bedrock. The telescope is an altitude-azimuth design (like a gun turret on a battleship). It addresses any point on the sky by tipping to the correct angle between zenith and horizon, and rotating around its vertical axis to the correct heading. The altitude bearing is comprised of a pair of giant C-rings that project the 500 tons of moving weight of the structure directly to the 150-ton azimuth frame and onto the pier. The C-ring bearings and azimuth bearings float on a film of oil pumped to 120 atmospheres pressure, allowing the telescope to be driven by four 3-horsepower motors on each axis.

The structure's towering center section provides the pivot points for swing arms that move auxiliary optics into and out of the main beams. Two swing arms support the prime focus cameras for direct imaging. They can be retracted and secondary mirrors put into place to form images at the straight-through Gregorian foci below the primary mirrors. A pair of flat tertiary mirrors can be swung into place to divert the light to bent Gregorian foci with the permanently mounted large instruments in the very center of the telescope structure. The primary mirrors will be protected by mirror covers that swing into place (just below the tertiaries) and that open like an oriental fan around a central hub.

Adaptive Optics

Modern large telescopes achieve their sharpest imaging by passing the light through a system to compensate for the blur of the Earth's atmosphere. Such systems are called adaptive optics (AO), with the central element being a thin, rapidly deformable mirror. Turbulent layers in the atmosphere high above the telescope produce cells in the air with very slight temperature differences from the surroundings. Just as in the case of waves in water on the surface of a pond, light waves that start out as spheres emanating from a distant astronomical source have traveled so far that they are nearly perfect plane waves at the top of the atmosphere. The atmospheric cells advance or retard just slightly the propagating light wave, turning a plane into a lumpy bumpy surface, resulting in a blurred star image. The adaptive optics system samples the shape of that surface and commands the deformable mirror to assume the 'inverse' pattern of lumps and bumps, restoring the incoming light pattern to nearly perfect focus. The mirror surface changes shape typically 1000 times a second.

The unique aspect of LBT's AO system is that the deformable mirror is the secondary mirror of the telescope itself. Adaptive optics systems provide the widest field of view and best correction in the infrared. At those wavelengths, room temperature optics and stray radiation from warm surfaces provide unwanted background noise to an already faint signal from the sources to be observed. Conventional AO systems have small deformable mirrors in cameras with a number of warm lenses to shrink the large telescope beam down onto the mirror, then to expand it outward again to the scientific instrument. Having the telescope's secondary mirror serve as the AO deformable mirror avoids the introduction of substantial extra background noise.

The challenge is in the size of the adaptive mirror. The thin shell of glass that makes the deformable reflecting element is 91 cm (36 inches) in diameter and only 1.6 mm (0.063 inches) thick. It is controlled by 672 electromagnetic actuators, with magnets glued directly onto its back surface. The magnets and associated electrical sensors pass through holes in a glass reference plate. The sensors measure the exact distance between the metallized surfaces of the thin shell and the reference plate, allowing control of the shape of the mirror to accuracy measured in nanometers. For initial implementation, the atmospheric blur must be measured with respect to a relatively bright star in the field of view; the corrections inferred from the 'wavefront sensing' camera are computed in a custom-developed processor and commanded to the electromagnets. A prototype system with half the number of actuators is showing great promise in operation at the MMT Observatory on Mt. Hopkins in Arizona.

■ **Below**: *Thin glass shell for adaptive optics secondary mirror, seen from the back side, with the mask pattern for gluing actuator magnets.*

The Partners

The successful completion of a visionary project requires creative initiators, solid engineering to bring a concept to reality, and inspired and inspiring entrepreneurs to attract the substantial support needed. University of Arizona astronomers Roger Angel and Neville Woolf conceived the basic LBT approach in the early 1980s. An engineer, Warren Davison, at the university played an important role in conceptualizing the mechanical structure with its C rings. One of the authors (JH) has been involved in leadership of the project for over 20 years, from the earliest days of producing borosilicate honeycomb mirrors. Peter Strittmatter, the Director of Steward Observatory, has been a driving force in forming and leading the consortium that runs the Observatory. The not-for-profit corporation was formed in 1992, and construction began in 1996. The University of Arizona holds a one-quarter share of LBT on behalf of the State University system, and is providing the two primary mirrors and adaptive secondary mirror thin shells. Phil Hinz, a University of Arizona astronomer, is the principal investigator for 'LBTI', the mid-infrared interferometer which has been supported by NASA for precursor observations for the Terrestrial Planet Finder mission. The University of Arizona hosts the LBT Observatory as a division of its College of Science.

Italian astronomers have played a critical role in the development and technical advances of LBT. Piero Salinari of the Arcetri Astrophysical Observatory in Florence led a team of astronomers and engineers in the design of the telescope mount. The telescope mount structure was built and assembled at Ansaldo-Camozzi Energy Special Components in Milan, and completed in 2001. It was then disassembled and shipped to Houston, whence the pieces were trucked to the base camp in Arizona. Salinari and his Arcetri colleagues, Simone Esposito and Armando Riccardi, are leading the scientific and industrial partnership to develop the adaptive secondary mirror systems, along with Daniele Gallieni of ADS and Roberto Biasi of Microgate. The astronomers Franco Pacini and Giancarlo Setti were instrumental in exciting their colleagues and institutions about forming a partnership to produce LBT. Emanuele Giallongo and his colleagues at the Rome Observatory are producing the 36-Megapixel Large Binocular Cameras for prime focus imaging. INAF, the Italian government agency supporting all of astronomy research in the country holds a one-quarter partnership share.

The LBTB is a consortium of German astronomical research institutes that also holds a one-quarter share in the LBT. It includes institutes in Heidelberg, Bonn, Potsdam, and Garching. Through partnerships, they are producing a major suite of instrumentation for the telescope. Holger Mandel and Walter Seifert in Heidelberg are leading the team creating 'LUCIFER', the pair of infrared imagers / spectrographs to be mounted at the outside bent Gregorian foci. Tom Herbst in Heidelberg is the PI of 'LINC-NIRVANA', the ambitious near-infrared interferometer with three levels of adaptive optics correction: ground layer and turbulent layers at two heights above the site. Klaus Strassmeier in Potsdam is producing the high dispersion, high stability spectrograph and spectropolarimeter, 'PEPSI', that will be mounted in its own climate-controlled chamber in the pier of the telescope and fed by optical fibers from the telescope foci.

Arizona-based Research Corporation provided the support for a one-eighth share. That observing opportunity will be divided among Ohio State University, the University of Notre Dame, the University of Minnesota, and the University of Virginia. The last one-eighth share belongs to Ohio State. Darren DePoy is the OSU astronomer leading the team to produce 'MODS', the pair of wide-field optical spectrographs. Bruce Atwood and a team of OSU engineers developed the giant bell jar that is the portable aluminizing facility, mounted directly to the primary mirror cells on the telescope.

Many other astronomers and administrators are involved in the challenge of the technical and financial development of the LBTO. The construction of the Observatory is costing $119 million, plus additional commissioning effort.

Significance of First Light and Path Forward

The stunning first-light image was taken with one primary and one LBC prime focus camera, with an overall telescope control system that was just beginning to function. It marks the first step on the road to cutting-edge science with this unique facility.

The goal for 2006 is achievement of 'second light', the initial imaging with both LBC's, and the commissioning of the telescope as a binocular system to the point that it is ready for

prime focus imaging science. The next milestone will be initial commissioning of the Gregorian foci with a simple fixed secondary mirrir in fall of 2007, followed by commissioning of the spectrographs LUCIFER and MODS. LUCIFER is a cryogenic spectrograph and imager designed to work in the near infrared. MODS is a multi-object spectrograph that works from the ultraviolet to the near infrared. Both spectrographs will be able to determine red shifts, chemical composition and physical conditions in many astronomical objects, and in particular, provide valuable follow-up observations to high redshift galaxies identified by the Hubble Space Telescope.

■ ***Above**: LBT open under moonlight on Emerald Peak.*

With arrival of the first adaptive secondary mirror system anticipated in early 2008, we will then see the telescope's first diffraction-limited imaging. The interferometers are scheduled for installation in the second half of 2008. The full instrument suite of first generation instruments in planned to be in full operation in 2009. We invite you to follow the progress of this exciting project at **http://www.lbto.org.**

6

As NASA's Flagship GREAT OBSERVATORY Completes 15 Years in Orbit...

Here, *Ray Villard*, Hubble's Press Officer since the launch of the Hubble Space Telescope in April 1990, brings us the highlights of the year from that most magnificent of instruments orbiting 600 kilometers above our heads.

...What's NEW with HUBBLE?

IN 2005, the Hubble Space Telescope was called into service for making a remarkably diverse assortment of observations of the universe. Beside peering at the most distant galaxies in the universe, it looked much closer to home by prospecting for resources on the Moon; intently watching the collision of a comet and a manmade object; finding more moons around Uranus and Pluto. In between these extremes of nearest and farthest objects in the universe, Hubble probed the secrets to the birth of extrasolar planets; the largest stars in the galaxy; and stars swirling around a nearby supermassive black hole.

Plans are moving forward to refurbish Hubble later in the decade. The telescope will be outfitted with a powerful new camera and ultraviolet spectrometer that will further boost its discovery potential. Once refurbished with new batteries, gyroscopes and other equipment, HST could potentially continue observing up until the 2013 launch of its successor, the James Webb Space Telescope. The HST servicing mission is contingent upon the Space Shuttle's successful return to flight in 2006. Hubble is booked on the manifest of future Shuttle launches after the return to flight demonstration is deemed successful.

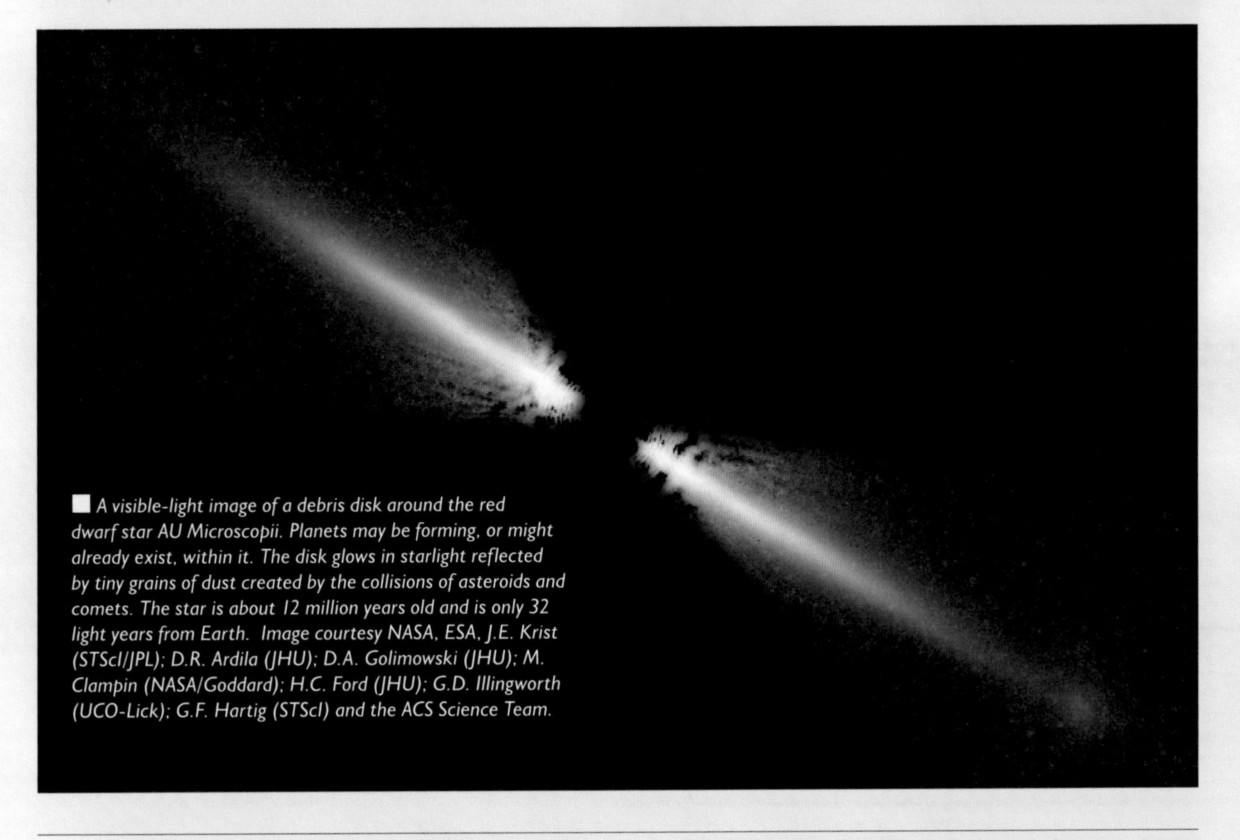

A visible-light image of a debris disk around the red dwarf star AU Microscopii. Planets may be forming, or might already exist, within it. The disk glows in starlight reflected by tiny grains of dust created by the collisions of asteroids and comets. The star is about 12 million years old and is only 32 light years from Earth. Image courtesy NASA, ESA, J.E. Krist (STScI/JPL); D.R. Ardila (JHU); D.A. Golimowski (JHU); M. Clampin (NASA/Goddard); H.C. Ford (JHU); G.D. Illingworth (UCO-Lick); G.F. Hartig (STScI) and the ACS Science Team.

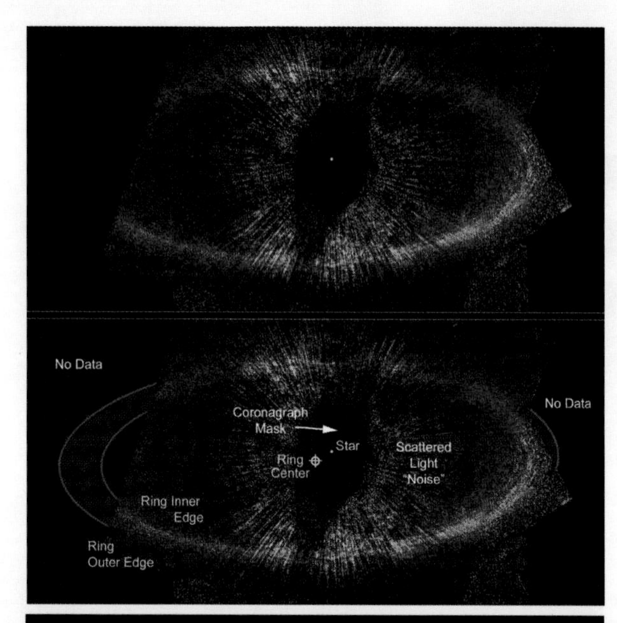

Hubble Observes the Star Fomalhaut

Hubble observations show that the bright star Fomalhaut is off-center from its surrounding ring of dust, most likely because of a planet orbiting the star and shepherding the ring material.

The star Fomalhaut

Center of offset dust ring

Hypothetical planet (estimated to be orbiting between 4.6 and 6.5 billion miles from the star)

Offset dust ring

Planetary Systems Under Construction

The discovery of nearly 200 planets around other stars doesn't offer many clues as to how planets form. But Hubble pictures provide the best look at the final stages of planet formation around other stars. Newborn planetary bodies condense out of a disk of dust and gas around newborn stars.

*■ **Top**: The top image is the first visible-light image of a dust ring around the nearby, bright young star Fomalhaut. The image offers the strongest evidence yet that an unruly planet may be tugging on the dusty belt. The left part of the ring is outside the telescope's view. The ring is tilted obliquely to our line of sight. The annotated image at bottom points out important features in the image, such as the ring's inner and outer edges. The Advanced Camera for Surveys' (ACS) coronagraph was used to block out the light from the bright star to make the ring visible. Image courtesy NASA, ESA, P. Kalas and J. Graham (University of California, Berkeley), and M. Clampin (NASA's Goddard Space Flight Center)*

*■ **Above**: Hubble observations show that the bright star Fomalhaut is off-center from its surrounding ring of dust, most likely because of a planet orbiting the star and shepherding the ring material. Illustration courtesy NASA, ESA and A. Field (STScI).*

These disks dissipate when the star ignites. But a second-generation dust disk forms as the young planetary bodies bump and grind as they collide with each other. These collisions pour dust into space forming the thin disk of around the star. This also happened in our Solar System – the collision that formed the Earth-Moon system being the most dramatic example – 4.4 billion years ago. By looking at dusty disks around other stars we are looking back into time and seeing what the early Solar System was like.

Observations by Hubble homed in on dusty disks around nearby stars. This at last offers snapshot of the entire sequence by which planets coagulate and gravitationally sweep out the interiors of the disks. Using Hubble, astronomers photographed thick disks of dust around a pair of young stars that might still be in the process of forming planets.

Hubble captured light reflected off a striking edge-on disk around a red dwarf star AU Microscopii. Hubble detected a cleared-out space in the middle as big as the orbit of Uranus. The Hubble images are so good they hints of warping due perhaps to the gravitational pull of a planet.

One of the most dramatic relics of planet birth is a vast dusty ring encircling the bright star Fomalhaut. This 200-million-year-old star lies 25 light-years from the Sun and is the brightest star in the constellation Piscis Austrinus. Though infrared observations over the past 20 years have inferred the presence of the ring, Hubble provides the first crisp infrared light picture of the ring and revealed that it's off center from the star.

This misalignment can best be explained by the gravitational pull of one or more unseen planets. It follows that the ring is skinny, rather than being a broad flattened disk, because a planet sweeps up material in its vicinity and also blocks dust from spreading inward toward the star. This conclusion is bolstered by the identification of a sharp inner edge to the ring, indicating a planet is sweeping out material. The suspected planet may be orbiting far away from Fomalhaut, near the dust ring's inner edge, between 7.6 billion and 10.5 billion kilometers (50 to 70 astronomical units) from the star.

First Image of An Extrasolar Planet?

Hubble's infrared camera provided one of the earliest images of what some astronomers consider the first extrasolar planet ever photographed. It is a companion to a relatively bright young brown dwarf star located 225 light-years away in the southern constellation Hydra. The European Southern Observatory's Very Large Telescope (VLT) in Chile made complementary observations of the planet candidate in April 2004. The follow-on Hubble infrared images were compared to the earlier VLT observations to try and see if the two objects are really gravitationally bound and move against the background stars together. The planet is so far from the dwarf (at least 30 percent farther apart than Pluto is from the Sun) that it would take 2,500 years to complete one orbit. Therefore the sluggish orbital motion could not be measured. Instead, by spring of 2005, astronomers saw enough of a shift against background stars that they were convinced the dwarf and planet are gravitationally linked together.

The infrared colors as measured by Hubble match theoretical expectations for an approximately 8 million-year-old object that is about five times as massive as Jupiter.

One lingering issue is the fact that the brown dwarf is only 25 Jupiter masses. So its companion object is one-fifth the dwarf's mass. By contrast, our entire Solar System of planets is less than one percent of the Sun's mass. So, could this really be classified as a binary brown dwarf system instead? This underscores the uncertainty in the semantics of what can and cannot be called a planet around another star.

A Big Galaxy in the Young Universe

Hubble's farthest-looking view of the universe, called the Hubble Ultra Deep Field (HUDF), revealed an unusually massive galaxy that may be among the most distant ever seen from Earth. Because the galaxy's light has taken approximately 13 billion years to reach us, we are seeing the galaxy as it formed just 800 million years after the Big Bang. Astronomers say the new galaxy is surprisingly massive and mature for its early age. This may raise questions about how galaxies are formed. Although the galaxy existed when the universe was only 6 percent of its present age, the galaxy has already built up to be as big as eight times the mass of our Milky Way. The galaxy could contain over 10 trillion stars. The discovery implies that star formation and the processes that led to the collapse of galaxies probably occurred much sooner after the universe cooled from its fiery big bang birth than once thought. Most galaxies seen at the universe's early age are generally much smaller.

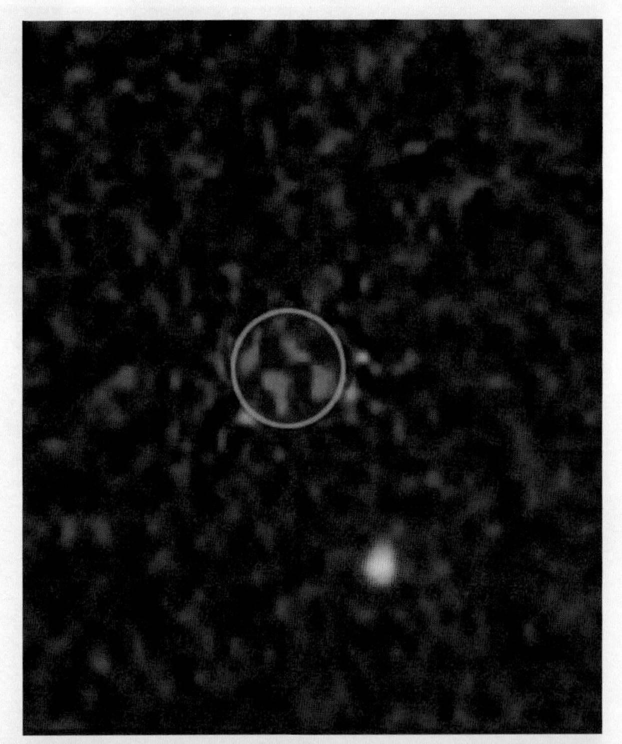

***Top**: This is an artificial-color Hubble Near Infrared Camera and Multi-Object Spectrometer (NICMOS) infrared-light view of the brown dwarf star 2MASSWJ 1207334-393254 (aka 2M1207) and giant planet companion candidate. The possible companion, estimated to be about five times the mass of Jupiter, is the magenta colored spot at lower right. The dwarf and candidate planet are at a minimum distance of 5 billion kilometers apart. At a temperature of only 1,800 degrees Fahrenheit, the candidate companion object appears very red in the NICMOS images. Image courtesy NASA, ESA, G. Schneider (University of Arizona).*

***Above**: Composite of visible-light (Hubble) and infrared (Spitzer) images of the distant galaxy HUDF-JD2 in the Hubble Ultra Deep Field. The faint red object was not visible in the Hubble image, but shows clearly in the infrared view. Image courtesy NASA, ESA and R. Hurt (Spitzer Science Center).*

This is interpreted as meaning that many of these smaller galaxies gradually combined over time to build larger galaxies like the Milky Way, in a process called the hierarchical growth of galaxies. The HUDF exquisitely chronicles the growth of galaxies from essentially super-star clusters, to tadpole-shaped merging systems, to fully-formed elliptical and spiral galaxies. To keep this paradigm intact, astronomers might shrug off this galaxy as just extreme example of a few galaxies that grew extremely quickly. Alternately, it may present a new challenge to galaxy formation theories.

Starry Ring around a Black Hole

Hubble identified the source of a mysterious blue light surrounding a supermassive black hole in our neighboring Andromeda Galaxy, M31. The strange light puzzled astronomers for more than a decade. As far back as 1995, Hubble revealed a strange blue light in Andromeda's core that astronomers said might have come from a single, bright blue star, or perhaps from a more exotic energetic process. Improved Hubble images show that the blue light is coming from a disk of hot, young stars. These stars are whipping around the black hole in much the same way as planets in our Solar System are revolving around the Sun. The blue light consists of more than 400 hot stars that formed in a burst of activity about 200 million years ago. The stars are tightly packed in a disk that is only one light-year across. The disk is

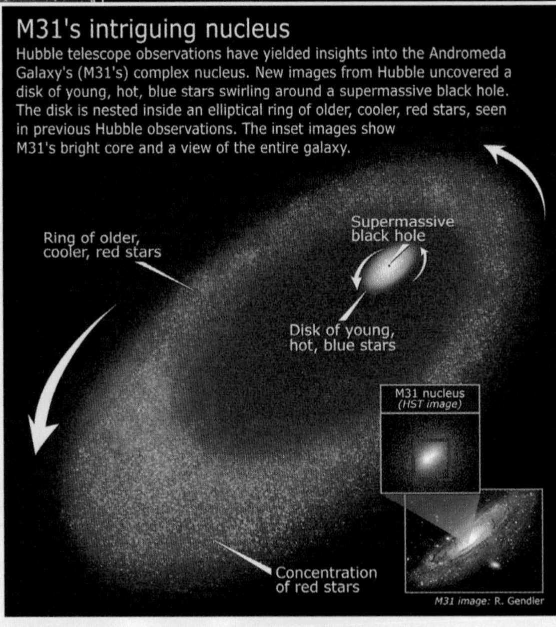

Top: *The Andromeda Galaxy (M31) on the left, with detail upper right showing the core is composed of a ring of old, red stars and a newly discovered disk of young, blue stars. The illustration at bottom right shows the structure of Andromeda's unusual core based on the upper image. The disk of blue stars is nested inside the larger ring of red stars. The tiny black dot within the blue disk is the monster black hole. Image at left courtesy Robert Gendler. Image at upper right courtesy NASA, ESA and T. Lauer (NOAO/AURA/NSF). Illustration lower right courtesy NASA, ESA and A. Feild (STScI).*

Right: *Hubble telescope observations have yielded insights into the Andromeda Galaxy's (M31's) complex nucleus. New images from Hubble uncovered a disk of young, hot, blue stars swirling around a supermassive black hole. The disk is nested inside an elliptical ring of older, cooler, red stars, seen in previous Hubble observations. The inset images show M31's bright core and a view of the entire galaxy. Image courtesy NASA, ESA, and A. Feild (STScI).*

of 140 million Suns, which is three times more massive than once thought. The evidence has helped astronomers rule out unlikely but alternative theories for the concentration of unseen mass in Andromeda's core.

The Largest Stars in the Galaxy

Using Hubble's infrared vision to peer deep into the core of our galaxy, astronomer's found an upper limit on the size of stars in the universe. The maximum mass is approximately 150 times the Sun's mass. To understand star formation and evolution astronomers need to know what factors determine the sizes of the most massive stars. Previous studies have shown that the smallest stars are about 8 percent our Sun's mass. But nailing down the mass of the largest star has been elusive. The universe's biggest stars have been of particular interest in recent years because of what happens when they die. Intense flashes of energy, called gamma ray bursts, are believed to be created with the collapse of the most massive of stars.

Researchers used Hubble for analyzing and measuring stars in the Arches Cluster, a spectacular hotbed of star formation near the galactic core where huge clouds of gas collide to form some of the largest stars known. The astronomers measured about 1,000 stars and found they ranged in size from six times the mass of the Sun to 130 times its mass. The researchers say that this is a conservative estimate and that the biggest stars in the cluster could be as much as 150 times the Sun's mass. The findings are consistent with star measurements taken of nine other clusters in earlier studies. The Arches cluster is an ideal target because it's made up of stars about 2.5 million years old, making them

nested inside a ring of older, redder stars, seen in previous Hubble observations.

Astronomers are perplexed about how the pancake-shaped disk of stars could form so close to a giant black hole. The conventional wisdom is that in such a hostile environment, the black hole's gravitational tidal forces should tear matter apart, making it difficult for gas to collapse and form stars. Based on X-ray observations of a disk-like structure in the center of our Milky Way, other astronomers propose that the gravity of the dense disk of gas is strong enough to offset the black hole's distorting tidal forces. With the two forces in balance, gas clouds can settle in and form stars.

The disk of stars also provides some of the best evidence to date for the existence of the supermassive black hole in M31. Astronomers used Hubble to measure the velocities of those stars. They obtained the stars' speeds by calculating how much their light waves are stretched and compressed as they travel around the black hole. Under the black hole's gravitational grip, the stars are traveling very fast: 3.5 million kilometers an hour. These velocities yield a precise estimate of the black hole's mass

Above: *This artist's impression shows how the Arches star cluster appears from deep inside the hub of our Milky Way galaxy. Although hidden from our direct view, the massive cluster lies 25,000 light-years away and is the densest known gathering of young stars in our galaxy. The illustration is based on infrared observations with Hubble and with ground-based telescopes, which pierced our galaxy's dusty core and snapped images of the luminous cluster of about 2,000 stars. Image courtesy NASA, ESA and A. Schaller (for STScI).*

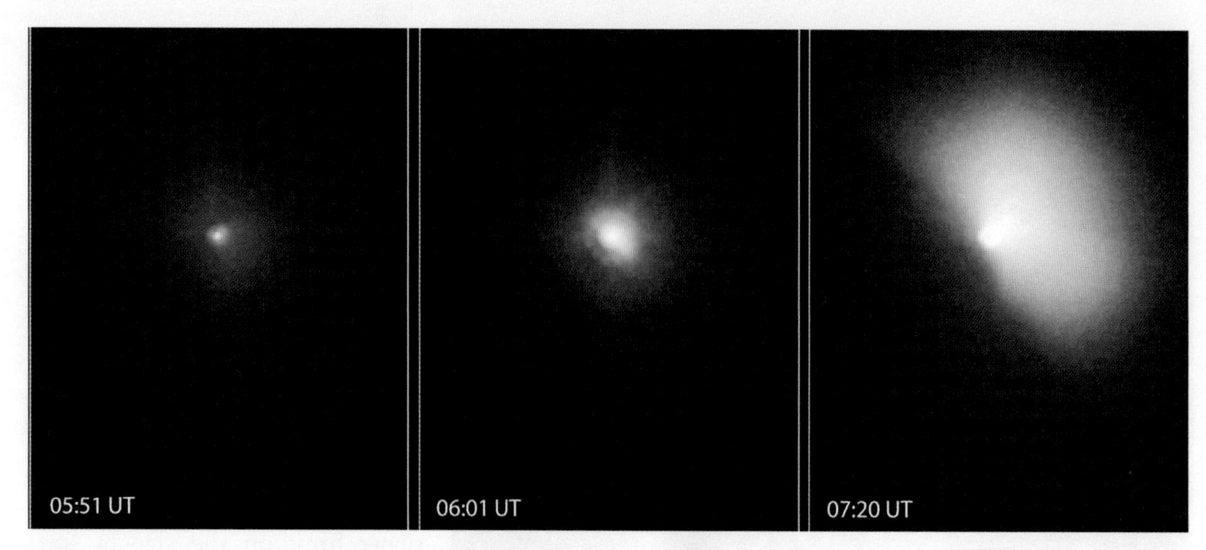

the right age. Stars that are in younger clusters are still shrouded in dust. In clusters of stars older than 2.5 million years, the most massive stars would already have exploded.

Comet Bull's-Eye

Hubble played an important supporting role in giving astronomers an early look at the collision of NASA's Deep Impact penetrator spacecraft with comet Tempel 2. On 4 July, the Deep Impact spacecraft slammed into the icy comet nucleus at 37,000 kilometers an hour, unleashing a force equivalent to about five tonnes of TNT. The resulting spray of dusty debris which rapidly fanned out from the nucleus was easily photographed by Hubble's sharp view. The dust quickly dissipated and no follow-on jets or outgassing, as predicted by some astronomers, were observed by Hubble. Analysis of the images from Hubble, other telescopes, and the Deep Impact mothership lead to the conclusion that the plume was a mix of ultra fine dust, like talcum powder, mixed with small crystals of ice. However Hubble saw no increase in emissions of water vapor or carbon dioxide, as anticipated. Hubble's spectrograph did measured traces of

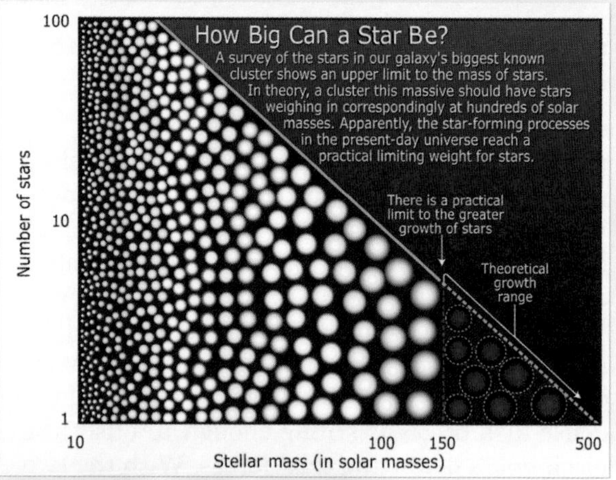

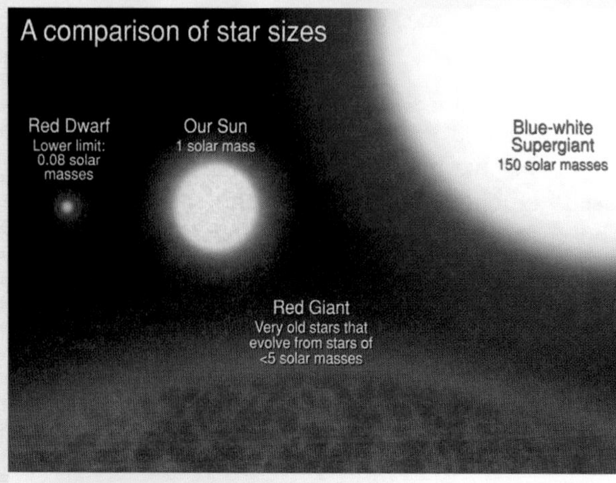

Above: *The dramatic effects of the collision early 4 July 2005 between a 370-kilogram projectile released by the Deep Impact spacecraft and comet 9P/Tempel 1. These Hubble images show the comet before and after the impact. The image at left shows the comet about a minute before the impact. In the middle image, 15 minutes after the collision, Tempel 1 appears four times brighter than in the pre-impact view. In the right image, taken 62 minutes after the encounter, gas and dust ejected during the impact are expanding outward in the shape of a fan. The fan-shaped debris is traveling at about 720 kilometers an hour. Images courtesy NASA, ESA, P. Feldman (Johns Hopkins University), and H. Weaver (Johns Hopkins University Applied Physics Lab.)*

Right/upper: *Unlike humans, stars are born with all the mass they will ever have. A star's mass ranges from less than a tenth to more than 100 times the mass of our Sun. Now astronomers have taken an important step toward establishing an upper mass limit for stars. Studying the densest known cluster of stars in our galaxy, the Arches cluster, astronomers determined that stars are not created any larger than about 150 times the mass of our Sun. Illustration courtesy NASA, ESA and A. Feild (STScI)*

Right/lower: *This illustration compares the different masses of stars. The lightest-weight stars are red dwarfs. They can be as small as one-twelfth the mass of our Sun. The heaviest-weight stars are blue-white super giants. They may get as large as 150 solar masses. Our Sun is between the lightweight and heavyweight stars. The red giant star at the bottom of the graphic is much larger than the other stars in the illustration. Its mass, however, can range from a fraction of the Sun's mass to a few solar masses. A red giant is a bloated star near the end of its life. In this brief phase, a star's diameter expands to several times its normal girth. Image courtesy NASA, ESA and A. Feild (STScI)*

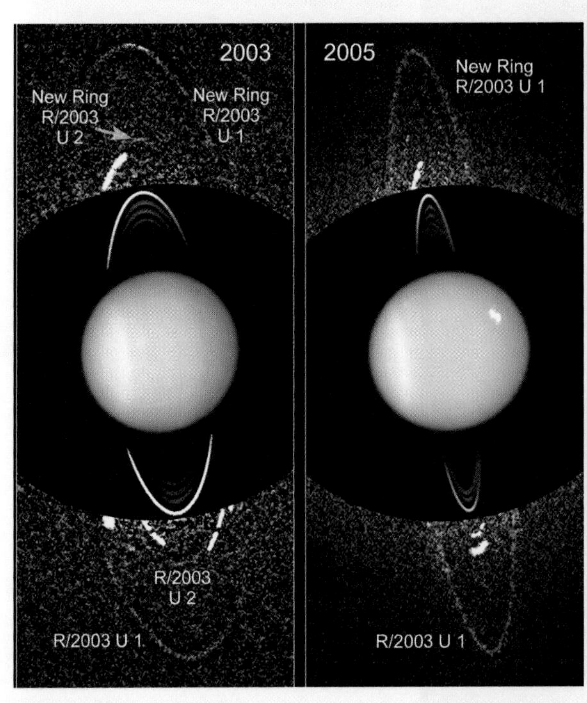

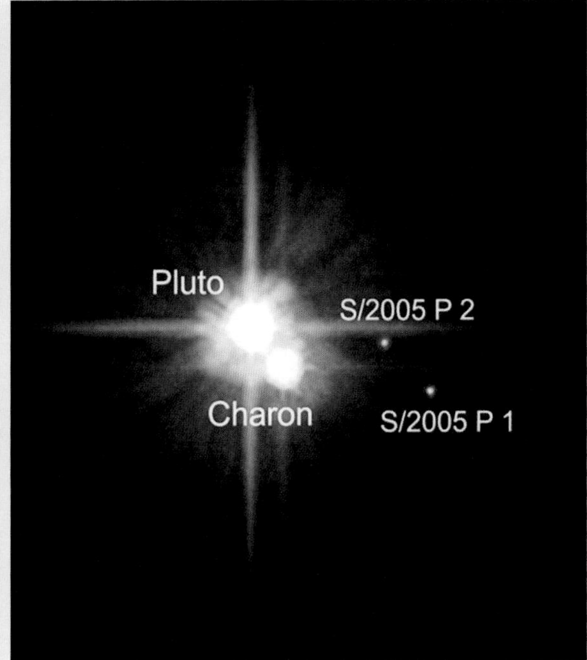

ethane and methanol. These two carbon-based molecules are among the chemical precursors to life on Earth.

New Moon and Rings around Uranus

Hubble discovered two tiny new moons and two faint new rings whirling around the planet Uranus. These escaped detection by the Voyager 2 space probe that flew by Uranus 20 years ago. The outermost new ring is twice the diameter of the previously known ring system. The finding raises the inventory for Uranus to 27 moons and 13 rings. The Hubble images reveal instabilities in the Uranian system – moons shifting orbit -- that could eventually lead to collisions between moons. This means that ever since its formation 4.5 billion years ago the Uranus system has been rapidly evolving. It is a dramatic demonstration that the processes which led to the formation of the planets are still ongoing today. The Hubble images revealed that one of those new rings is associated with Mab, one of the new moons. The dust that forms the ring is continuously blasted off Mab by meteorite collisions with the moon escapes into orbit and replenishes the ring.

New Moons for Pluto

Hubble discovered that Pluto has two additional tiny moons whirling around it. Pluto's only previously known moon, Charon, was discovered in 1978, almost a half century after Pluto's discovery in 1930. The newfound satellites make Pluto a unique "quadruple" system of co-orbiting bodies, and therefore more complicated than the binary asteroids and binary Kuiper Belt objects being discovered. Depending on their reflectivity, the new satellites are only tens of miles across, and so they're roughly 1/5,000th Pluto's brightness. This explains why they had not been seen previously. The new moons, now named Nix and Hydra, are roughly two to three times as far from Pluto as its large moon Charon.

The satellites appear to be in circular orbits and co-planar with the orbit of Charon. This strongly suggests that the new moons were born from a giant collision between Pluto and a similarly sized body over 4 billion years ago. Some of the debris from the impact which went into orbit around Pluto and formed Charon, also produced other moons as well. Earth's Moon was born through a similar collision, so further study of Pluto may lead to a better understanding of the origin and evolution of the Earth-Moon system. The newly-discovered moons are constantly being hit by space debris which should generate rings around Pluto. So far, no one has seen them if they are there. But such rings could pose a navigational hazard to NASA's New Horizons spacecraft which will visit Pluto in 2015.

■ ***Top Left**: These composite images from several observations by NASA's Hubble Space Telescope reveal a pair of newly discovered rings encircling the planet Uranus. The left composite image is made from Hubble data taken in 2003. The new dusty rings are extremely faint and required long exposures to capture their image. In the image at right, taken two years later, the rings appear more oblique because Uranus has moved along its solar orbit. Images courtesy NASA, ESA, and M. Showalter (SETI Institute)*

■ ***Top Right**: Pluto, Charon, and the two new moons discovered in this image taken by the Hubble Space Telescope. Image courtesy NASA, ESA, H. Weaver (JHU/APL), A. Stern (SwRI), and the HST Pluto Companion Search Team.*

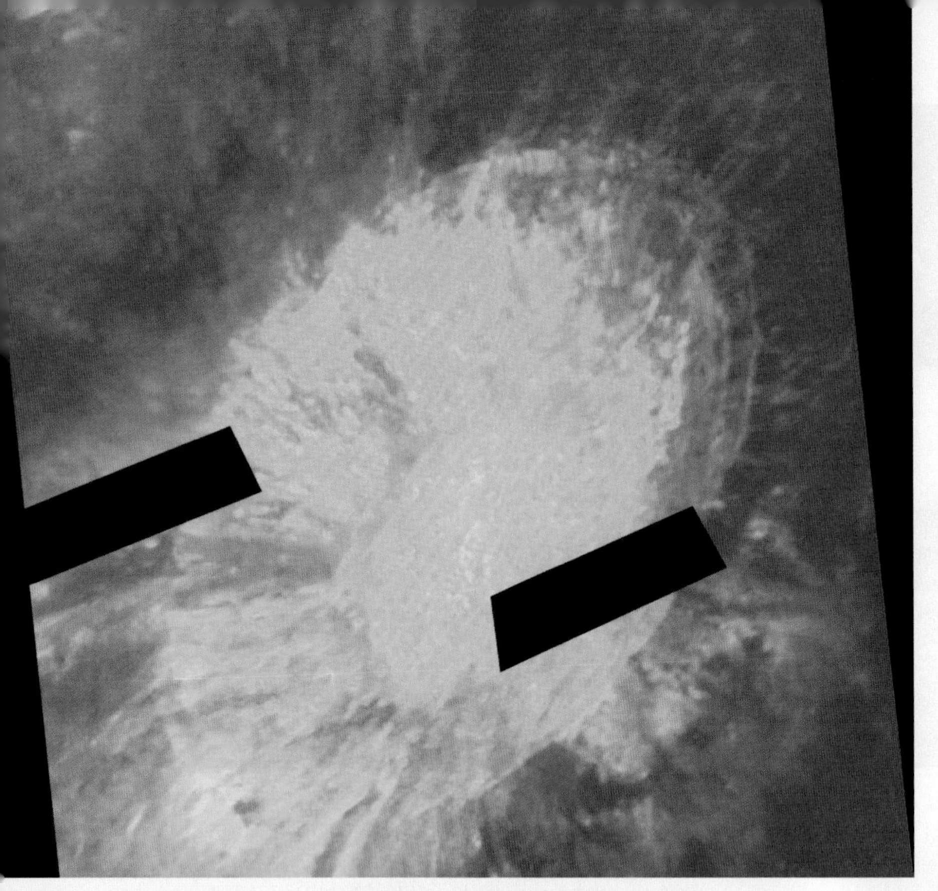

Lunar Prospecting

Hubble was called into service to support NASA's plans to send humans to the Moon and establish outposts. Hubble served as a lunar prospector, looking for valuable minerals that would provide resources for a lunar base. Besides water-ice, which has been inferred from lunar orbiting probes, scientists are also looking for elements that can provide precious oxygen for air and propellant.

Hubble took ultraviolet and visible-light images of the geologically diverse Aristarchus impact crater and the adjacent Schröter's Valley rille, an ancient lava flow. Hubble also photographed the Apollo 15 and 17 landing sites in UV light. (The artifacts left on the Moon surface by the Apollo visits are too small to be seen by Hubble.) Astronauts collected and returned from the Moon rock and soil samples in 1971 and 1972. These specimens provide "ground truth" for comparing the UV appearance of known minerals returned form the Moon to the unknown minerals in the Aristarchus region.

The Aristarchus plateau is unique for its volcanic vents, collapsed lava tubes, ejected volcanic material and recent impact craters. The Aristarchus crater, 42 kilometers wide and three kilometers deep, could be as young as 100 million years old. The crater's sharp rim and other fresh features reveal the varied geology of the plateau. The crater slices into the side of the plateau, exposing its interior layers and features.

As anticipated, Hubble's images showed a diversity of materials in the crater: basalt, olivine, anorthosite and ilmenite. Ilmenite, a glassy mineral made up of titanium, iron and oxygen, was identified in samples returned from the Apollo 17 site.

The telescope's findings pave the way for further prospecting by NASA's Lunar Reconnaissance Orbiter mission planned for 2008.

As Hubble completes its 15th year of operation, NASA's flagship Great Observatory is in the prime of its age of exploration. Hubble has far exceeded the expectations of astronomers when it was launched in 1990.

The Dwarf Planet Ceres

The largest of the asteroids (roughly the size of Texas), Ceres was the first asteroid to be discovered in 1801. For the past 200 years astronomers have tried to look closely at Ceres with ever larger telescopes. Finally, Hubble provided sharp enough images to determine that Ceres is quite likely a "mini-planet." The asteroid has enough gravity to make it almost spherical. Analysis also suggests that, like Earth, it may have different layers of increasingly denser material toward the center, like the layering in a parfait desert. If the asteroid was just a ball of equal density material, it would gravitationally settle into a more oblate shape -- flatter at the poles and bulging at the equator. This means there's something inside that keep Ceres from deforming. The likeliest material is water ice, which expands when it freezes. If that is so, Ceres may have more fresh water than all the fresh water on Earth. This would make it an invaluable source of raw material for the human exploration of the Solar System.

***Above**: This color composite focuses on the 42-kilometer diameter Aristarchus impact crater, and employs ultraviolet-to visible-color-ratio information to accentuate differences that are potentially diagnostic of ilmenite- (i.e. titanium oxide) bearing materials as well as pyroclastic glasses. The symphony of color within the Aristarchus crater clearly shows a diversity of materials — anorthosite, basalt, and olivine. The images were acquired 21 August 2005. The processing was accomplished by the Hubble Space Telescope Lunar Exploration Team at NASA's Goddard Space Flight Center, Northwestern University, and the Space Telescope Science Institute. False-color images were constructed using the red channel as 502/250 nanometers; the green as 502 nanometers; and the blue as 250/658 nanometers. North is at the top in the image.*

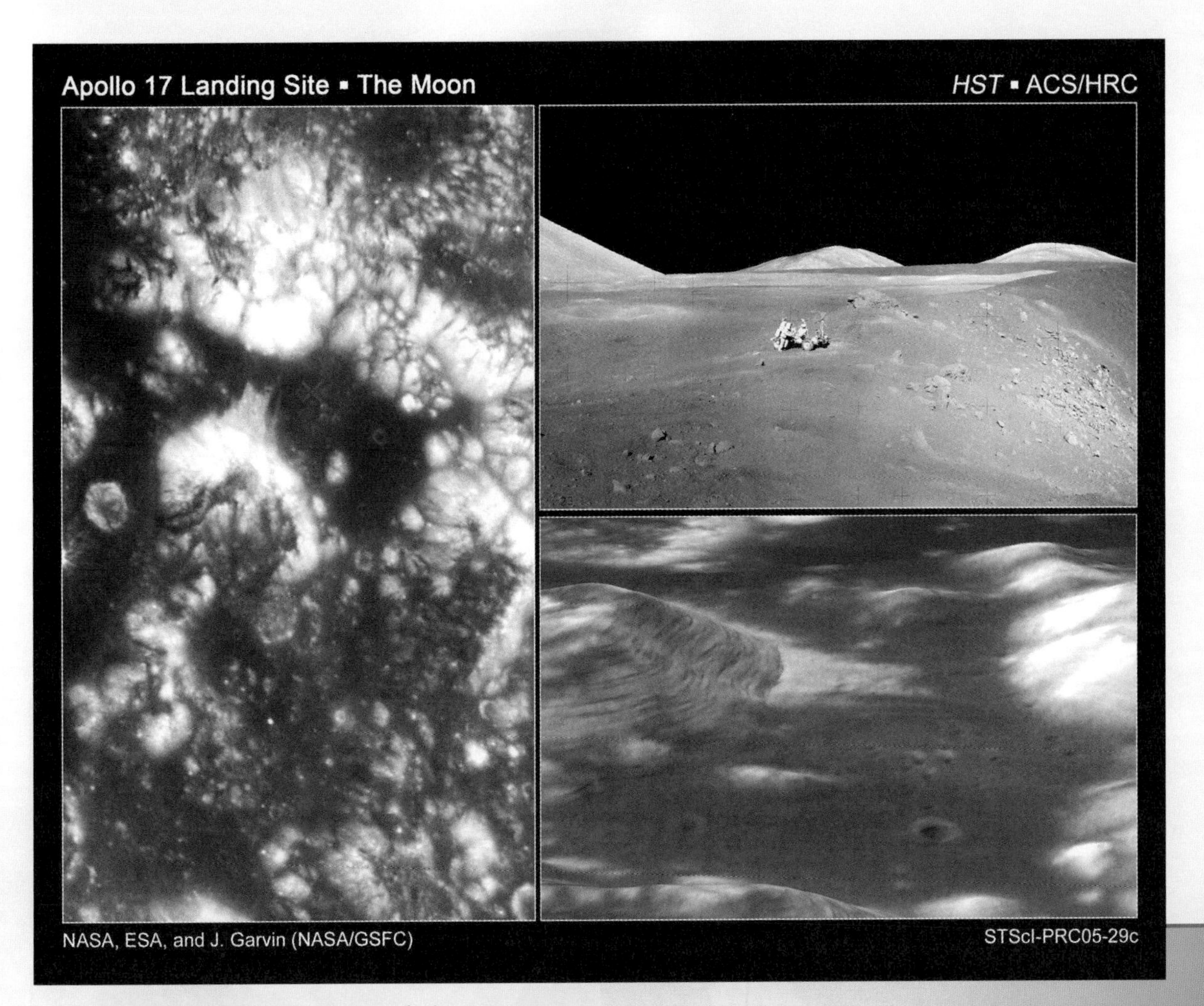

Peering to the far horizon of the universe, Hubble teamed up with NASA's other two Great Observatories, the infrared Spitzer Space Telescope and Chandra X-Ray Observatory, to discover and analyze newly forming galaxies that emerged when the universe was only one-tenth of its present age. Looking closer to home, Hubble set limits on the masses of the largest stars in our galaxy, as well as precisely measure the mass of the smallest star, such a the white dwarf orbiting the winter star Sirius. Some of Hubble's most remarkable findings were in our own cosmic backyard. Hubble's discovery of new moons around Pluto, and more rings around Uranus dramatically demonstrate that more surprises are awaiting us among the planets.

Now Hubble is even playing a role in planning for the return of astronauts to the Moon by spectroscopically identifying minerals crucial to sustaining a human outpost on the Moon. Considering these remarkable advances, it is hard to imagine what astronomy would have been like if instead Hubble was retired at 15 years, as originally planned. Instead, some of Hubble's most exciting discoveries have yet to be made.

Above: *This image showcases Hubble Space Telescope's first high-resolution ultraviolet and visible imaging of the Apollo 17 landing region within the Taurus-Littrow valley of the Moon. Humans last walked and drove on the lunar surface in this region in December 1972. The image at upper right was taken by the Apollo 17 astronauts (Dr. Harrison Schmitt and Gene Cernan). It illustrates a view of the rim of Shorty crater and the lunar roving vehicle against a backdrop of the mountain-like massifs that define the Taurus-Littrow valley. The Hubble Space Telescope image at lower right was constructed by overlaying the Hubble image (at left) with a digital-terrain model acquired by the Apollo program to provide a perspective view looking from west to east up the valley. Image courtesy NASA, ESA, and J. Garvin (NASA/GSFC).*

Hubble continues to search the far universe for the most distant galaxies. New gravitational lenses yield data about the expanse of the enigmatic dark matter. Peering into the heart of globular clusters, or into the heart of galaxies, Hubble's sharp eye reveals activity in the densest parts of our universe. Closer to home, we can expect new discoveries in star formation and planet formation. Hubble's future, including a possible new suite of instruments, is bright indeed.

7
More than just OUR GALAXY...

Astronomers are making dramatic discoveries about the galaxy in which we live. In this chapter *Christopher Wanjek*, an experienced science writer, reviews the latest results, from the galactic center, to the central bar, and way out to its warped disk.

...The New MILKY WAY

A FLIGHT THROUGH our Milky Way galaxy, could you afford faster-than-light travel, would reveal over 200 billion stars, including millions of black holes and neutron stars thrown in for a little excitement. The voyage would take you through a multitude of stellar nurseries, where cold and massive gas clouds collapse to form to stars, and onward to new solar systems, likely peppered with earthlike planets. The one-way trip across the galaxy and all four major spiral arms stretches for about 100,000 light-years.

From a more modest vantage point here on Earth, atop mountain observatories and with orbiting telescopes only a handful of miles above these, astronomers have made a series of remarkable discoveries about our home galaxy in the past year. It seems that we have miscalculated the very shape and contents of our galaxy.

Star Death

One study smacks of forensic detective work, a 'CSI: Milky Way'. Astronomers measured the rate of star death and rebirth in our galaxy by combing through the sparse remains of exploded stars from the last few million years. As reported in Nature, a team led by Roland Diehl of Max Planck Institute for Extraterrestrial Physics in Garching, Germany, used the European Space Agency's INTEGRAL satellite to explore regions of the galaxy shining brightly from the radioactive decay of ^{26}Al, an isotope of aluminum. This aluminum is produced in massive stars and in their explosions, called supernovae, and it emits a telltale light signal in the gamma-ray energy range.

The team's multi-year analysis revealed three key findings: ^{26}Al is found primarily in

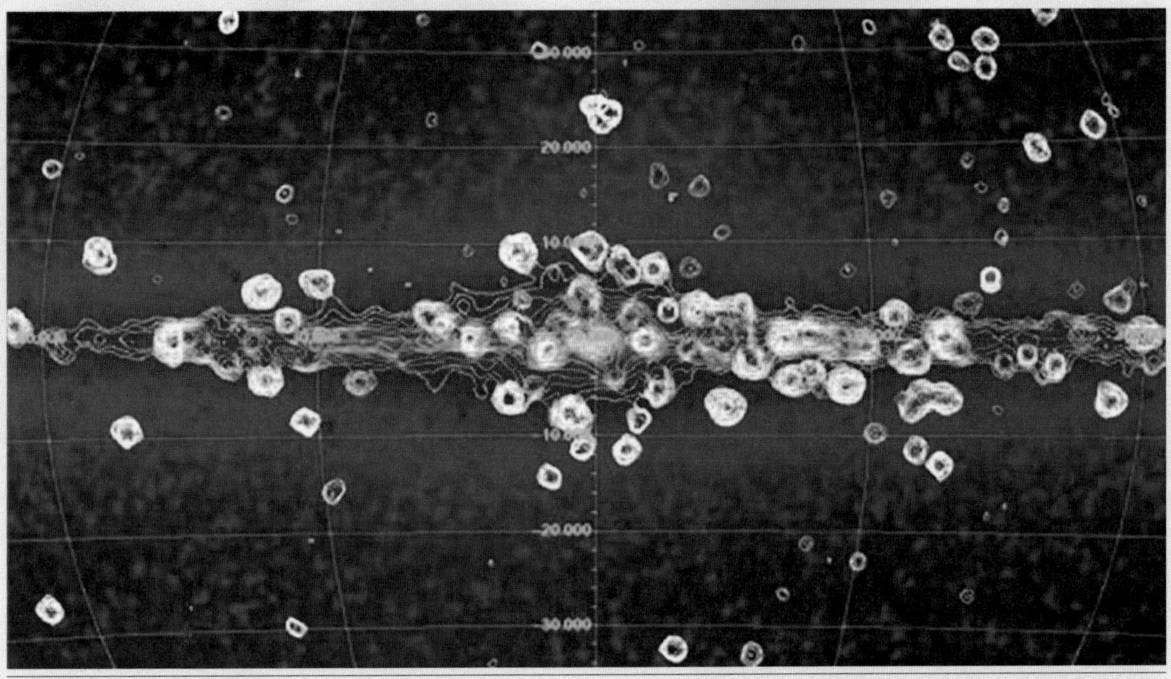

star-forming regions throughout the galaxy; about once every 50 years a massive star will go supernova in our galaxy (yes, we are overdue); and each year our galaxy creates on average about seven new stars. So, while our galaxy isn't the biggest producer of stars and supernovae in the universe, there's still plenty of activity.

And there is indeed much activity out there we don't yet know about. Nearly 400 years ago, Galileo determined the wispy Milky Way actually comprises myriad individual stars. Scientists using NASA's Rossi X-ray Timing Explorer now have done the same for the "X-ray" Milky Way. The origin of this X-ray counterpart to the Milky Way, known to scientists as the galactic X-ray background, has been a long-standing mystery. A team led by Mikhail Revnivtsev of the Max Planck Institute for Astrophysics, also in Garching, has determined that the background is not diffuse, as many have thought. Rather, it emanates from untold hundreds of millions of individual sources dominated by a type of dead star called a white dwarf, along with stars with unusually strong coronas.

The result solves major theoretical problems yet points to a surprising undercounting of stellar objects, perhaps by a hundredfold. The white dwarfs are locked in binary systems called cataclysmic variables, in which a normal sun-like star spills gas towards the white dwarf to create X-rays. Compared to their denser and meaner cousins—black holes and neutron stars—white dwarfs elicit yawns from many astronomers. But these underdogs, due to their sheer number, dominate the high-energy radiation output in the Milky Way, a surprise and humbling finding.

The Warped Disk and a Visitor

For a half century scientists could not explain the warp seen in the thin disk of our galaxy. This is a slight bending in the disk of neutral hydrogen gas. Like a vinyl record that has seen better days, the warp causes the galaxy to wobble slightly as it rotates. The culprit may be our neighbors, the Large and Small Magellanic Clouds. Leo Blitz of the University of California, Berkeley, led a thorough analysis of the hydrogen disk based on optical data, and he plugged this into a computer simulation by Martin Weinberg of the University of Massachusetts, Amherst. These two satellite galaxies were long considered too small to disturb our Milky Way; their combined mass is only 2 percent of our galaxy's mass. Yet there's more than meets the eye. Computer simulations show that the two galaxies are sailing through a sea of the Milky Way's unseen dark matter. The wake causes the Milky Way disk to oscillate at the observed frequencies.

The Milky Way apparently has a new visitor, too, a newly detected dwarf galaxy that our galaxy is now consuming. This discovery was made through the Sloan Digital Sky Survey, an ambitious optical survey to precisely map one-fourth of the entire sky. The survey has already mapped a mind-boggling 48 million stars. Robert Lupton of Princeton University, part of the discovery team, compared the Sloan survey

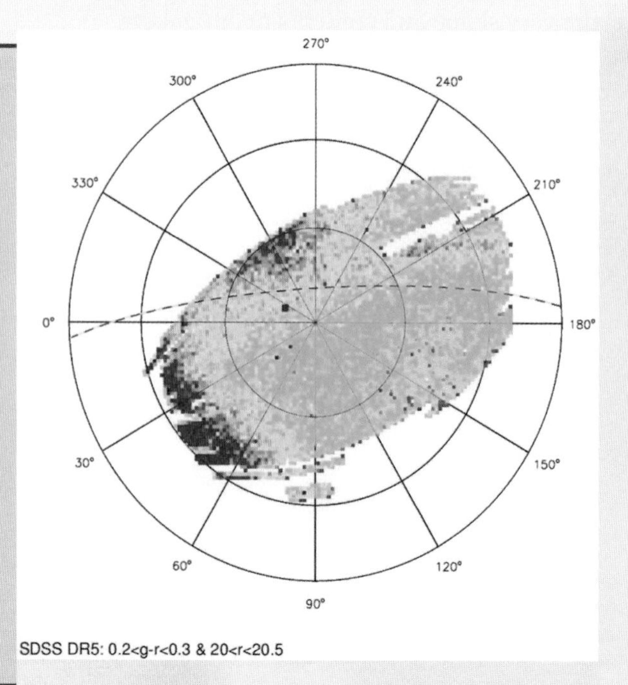

■ ***Previous page**: An image of the galactic X-ray background superimposed on an image of infrared sources. The X-rays, shown as white contour lines, were detected by NASA's Rossi X-ray Timing Explorer. The infrared background (everything else) was detected by NASA's COBE mission in the 1990s. The white knots reveal very bright X-ray sources, mostly from black hole and neutron star activity. Most of the X-rays in our galaxy, however, come from millions of relatively dim sources that cannot be imaged. These are the thin, white, wavy lines throughout the galaxy. These appear to be almost entirely unseen white dwarfs and stars with active coronas, a surprising discovery. Image courtesy NASA/RXTE-COBE/Revnivtsev et al.*

■ ***Above**: A huge but very faint structure, containing hundreds of thousands of faint blue stars, has been discovered by astronomers of the Sloan Digital Sky Survey (SDSS-II). At an estimated distance of 30,000 light years (10 kpc) from Earth, the structure does not follow any of Milky Way's three main components: a flattened disk of stars in which the sun resides, a bulge of stars at the center of the galaxy and an extended, roughly spherical, stellar halo. The most likely interpretation of the new structure is a dwarf galaxy that is merging into the Milky Way. Illustration courtesy Mario Juric (Princeton U.)/SDSS-II Collaboration.*

to wearing a pair of 3-d glasses. Structures once lost in the background snap into view. The tiny galaxy contains hundreds of thousands of stars and is relatively close to Earth, only 30,000 light-years away. This is so close, in fact, that astronomers assumed the galaxy's stars were ordinary Milky Way stars. The Sloan survey revealed how the galaxy is a unique entity.

Mapping the Galaxy

The problem with mapping the Milky Way is that it is big and we see it in two dimensions. Large optical surveys such as the Sloan Digital Sky Survey help. Sharp vision is also key. This is where radio astronomy comes in. Vast arrays of interconnected radio dishes, such as the Very Large Array (VLA) in New Mexico and the Very Long Baseline Array (VLBA) strung from Hawaii to the Caribbean, provide resolution in radio frequencies that can surpass what the mighty Hubble Space Telescope can do in optical. The VLA and VLBA, as well as a good deal of other ground-based astronomy, are funded by the National Science Foundation.

Mark Reid of the Harvard-Smithsonian Center for Astrophysics in Cambridge, Mass., used the VLBA to discover that the Perseus spiral arm, the nearest spiral arm to ours, lies only half as far from Earth as some previous studies had suggested. We are located in the Orion arm, a spur of the major Sagittarius arm. The Perseus spiral arm is a mere 6,400 light-years. Reid's team measured the region around a newly formed star in Perseus with an accuracy of 10 micro-arcseconds. That resolution is equivalent to determining from Earth whether an astronaut on the Moon is holding a flashlight in his left or right hand.

NASA's Spitzer Space Telescope, which observes the infrared waveband, has provided a new view of the galaxy as a whole. Dust is largely transparent to infrared light, and Spitzer could ascertain a broad view of the galaxy through the dusty spiral arms in ways not possible in other wavelengths. The Milky Way, it seems, is no ordinary spiral galaxy. A team led by Ed Churchwell of the University of Wisconsin-Madison measured a long, central bar in the galaxy's center, at the origin of the spiral arms, about 27,000 light-years across, at least 7,000 light-years longer than predicted. The observation supports the notion that ours is a barred spiral galaxy, yet with a longer bar and more spiral arms than other barred spirals.

All Quiet – For Now...

Things are quiet in our spot of the galaxy, where the closest star is over four light-years away. This tranquility enabled life to gain a foothold on Earth. There is little threat of a supernova exploding nearby and wiping out our atmosphere with an influx of X-ray and gamma-ray radiation. Yet perhaps once during this history of life on Earth a star exploded close enough to cause a mass extinction. A team led by Adrian Melott of the University of Kansas speculates that a gamma-ray burst from a supernova event could have zapped Earth 450 million years ago to cause the Ordovician extinction, which killed off 60 percent of all marine invertebrates. The teams work is based on known, albeit sketchy, gamma-ray burst rates and impressive computer modeling that details the effect of gamma rays on the atmosphere. A 10-second gamma-ray burst originating within 6,000 light-years from Earth would have a devastating effect on life, wiping out half the ozone layer for a few weeks.

The Galactic Center

Life as we know it has little chance of thriving 26,000 light-years away in the dense galactic core, a beehive of activity where 20 percent of the Milky Way's stars reside. There is no escape from the searing radiation associated with the furious pace of star formation and star death. At the center of the galaxy lies Sagittarius A*,

■ **Above**: *The Perseus spiral arm - the nearest spiral arm in the Milky Way outside the Sun's orbit - lies only half as far from Earth as some previous studies had suggested, only 6,400 light-years away. An international team of astronomers measured a highly accurate distance to the Perseus arm for the first time using a globe-spanning system of radio dishes known as the Very Long Baseline Array (VLBA). Illustration courtesy Y. Xu et al.*

and gamma rays are lighting up the hydrogen gas in Sagittarius B2 today.

A team led by Zhi-Qiang Shen of the Shanghai Astronomical Observatory used the VLBA radio observatory to determine the size of Sagittarius A*. The region could fit between the Earth and the Sun, the team found. With the mass of millions of suns packed into such a tiny region, this is the best evidence to date that Sagittarius A* is indeed a black hole. Astronomers using NASA's Chandra X-ray Observatory documented a surprising tale of survival around Sagittarius A*: New stars are forming as close as a light-year from the black hole. The very matter falling into the black hole—largely from stars torn apart—can spur new star formation by providing the raw materials. The disk of black hole fuel extends many light-years away from Sagittarius A*. The gravity of the disk can offset the gravity from the black hole, enabling stars to form in between.

a suspected supermassive black hole containing the mass of about 4 million suns. Sagittarius A* isn't that active today but may be been as little as 350 years ago, according to new gamma-ray observations. Mikhail Revnivtsev, who was behind this year's X-ray background discovery, used the European Space Agency's INTEGRAL observatory to study a region 350 light-years away from the black hole, a region called Sagittarius B2. Here Revnivtsev found the region is being pelted with debris from chaotic black hole activity—more specifically, the X-rays and gamma rays emitted from the black hole region, the final light released from matter before being pulled into the void. Those X-rays

Above: *This artist's rendering shows a view of our own Milky Way galaxy and its central bar as it might appear if viewed from above. New observations by the GLIMPSE legacy team with NASA's Spitzer Space Telescope indicate that the bar-shaped collection of old stars at the center of our galaxy may be longer, and at a different orientation, than previously believed. Our Milky Way galaxy may appear to be very different from an ordinary spiral galaxy. Illustration courtesy NASA/JPL-Caltech/R. Hurt (SSC).*

Opposite: *A new infrared mosaic from NASA's Spitzer Space Telescope offers a stunning view of the stellar hustle and bustle that takes place at our Milky Way galaxy's center. The picture shows throngs of mostly old stars, on the order of hundreds of thousands, amid fantastically detailed clouds of glowing dust lit up by younger, massive stars. Image courtesy NASA/JPL-Caltech/S. Stolovy (SSC/Caltech).*

The Spitzer telescope, like the Chandra Observatory a few years ago, captured the entire galactic center in all its stunning glory. Like the Chandra X-ray image, the Spitzer infrared image is a mosaic of many observations. The new picture shows throngs of mostly old stars, on the order of hundreds of thousands, amid fantastically detailed clouds of glowing dust lit up by younger, massive stars. Features within the new mosaic include dust clouds of a dizzying variety, such as glowing filaments, wind-blown lobes flapping outward from the plane of the galaxy. The Spitzer image also shows newborn stars just beginning to break out of their dark and dusty cocoons.

Is there life out there? That's the million-dollar question. Spitzer added a little excitement to the search with the discovery of chemicals needed for life sprinkled throughout the galaxy—and beyond, for that matter. A team led by Douglas Hudgins of NASA Ames Research Center found an abundance of nitrogen-rich polycyclic aromatic hydrocarbons. Without the nitrogen, this is a largely cancer-causing chemical family that includes benzene. With the nitrogen, they are an important component of hemoglobin and chlorophyll, not to mention caffeine and chocolate. The existence of simple extraterrestrial life forms shouldn't be a surprise given the richness of life-yielding chemicals in the Milky Way.

Dark Matter and Anti-Matter

In the realm of the slightly more speculative was a science announcement about dark matter, one of the biggest mysteries in astronomy. The universe contains only 4 percent ordinary matter, or familiar atoms. Unidentified dark matter makes up 22 percent of the universe, and a mysterious dark energy driving an accelerated expansion of the universe fills out the remaining 74 percent.

Ongoing observations of the galactic center reveal copious amounts of antimatter. That's not so strange. Antimatter is regularly created in particle collisions in space, often associated with supernovae. When newly minted anti-electrons, or positrons, soon encounter electrons, they annihilate and produce a telltale gamma-ray photon at the 511 keV energy level. A team led by Georg Weidenspointner of the French laboratory for space physics, CESR, found too much antimatter to explain away with supernovae. Using INTEGRAL, his team found the antimatter concentrated in the galactic center. Its energy level and distribution fits with one theory of dark matter particle collisions. If confirmed this would be a landmark discovery, the first direct detection of dark matter. But confirmation will not come for a few more years, as much more data are needed.

How Old?

With all this activity in the Milky Way, you may have guessed that the galaxy has been around for a long time. Indeed, our galaxy was likely assembled within a billion years after the big bang, or about 12.5 billion years ago. This has long been a mere assumption, however, based on observations of the very first galaxies by observatories such as Hubble. The existence of local white dwarfs (which we now know dominate the X-ray background) also implies an age of at least 11 million years—the time needed for a star like our sun to form, run out of fuel, go nova, and collapse into a white dwarf. The University of Chicago's Nicolas Dauphas came close to the 12.5 billion number by dating real material in our galaxy. As discussed in Nature, Dauphas compared the decay of two long-lived radioactive elements, uranium-238 and thorium-232 in a very old halo star. The technique is akin to carbon dating with carbon-14 and carbon-12 ratios.

Dauphas' result actually places the age of the Milky Way at 14.5 billion plus or minus 2 billion. With NASA's WMAP mission precisely dating the universe as 13.7 billion years old, we can assume the Milky Way is a little younger. Nevertheless, Dauphas's work highlights the fact that the Milky Way is still full of surprises.

8

The Search for GRAVITATIONAL WAVES...

Bunny really enjoyed surfing Gravitational Waves: but they interrupted the Human Astronomers' Research!!

As the Laser Interferometer Gravitational-wave Observatory (LIGO) begins a continuous run of observations, *Laura Cadonati* of MIT gives us the benefit of her considerable expertise to tell the story of the remarkable set of instruments at both LIGO sites. There is the potential for great discoveries within the next few years.

...Listening to SPACE with LIGO

THE 2005 World Year of Physics was dedicated to Albert Einstein, on the centennial of the "miraculous year" when his breakthrough papers on Brownian motion, light quanta and special relativity were published, reshaping modern physics. The year 2005 also marked a major step towards the fulfillment of Einstein's vision, as scientists at the Laser Interferometer Gravitational-wave Observatory (LIGO) delivered high precision instruments that are now listening for gravitational waves, elusive distortions of space-time predicted by General Relativity. To understand the nature of gravitational waves and why an army of physicists and engineers is trying to detect them, it is useful to review how Einstein's theory of General Relativity redefined our understanding of the Universe.

From Newton to Einstein

Sir Isaac Newton's theory of universal gravitation (Principia Mathematica, 1687) was brilliant. Inspired by the works of Kepler and Galileo Galilei, Newton observed that any two objects are attracted to each other by a force proportional to their masses and inversely proportional to the square of the distance between them. This observation, combined with the laws of classical mechanics, explained several puzzling phenomena, like the orbits of planets and moons and the oceans' tides. Newton depicted a peaceful universe where all celestial bodies know about each other's presence and move in a harmonious balance of forces. In all of its elegance, though, Newton's theory has two fundamental problems. First, it does not explain what mechanism is behind the force of attraction between masses. Second, it declares that gravity performs "action at a distance": whenever something changes in the position of a celestial body, all other stars and planets instantly know about it and readjust their positions and orbits accordingly, with the uncomfortable consequence that information travels faster than light.

About 300 years passed before the open issues in Newton's model were addressed by General Relativity. According to Einstein, there are no invisible wires pulling bodies towards each other. What we experience as a force of attraction is rather a geometrical property of the four-dimensional universe we live in ("space-time"). To visualize this, a classical two-dimensional analogy can be drawn between space-time and a rubber sheet. A bowling ball on the sheet will deform it and a smaller object in its vicinity, like a golf ball, will move towards the larger mass not because it is pulled by some invisible force, but simply because it is sitting on a curved surface. In the four-dimensional extension of this picture, Earth's orbit is a geodesic in the space-time deformed by the Sun, clocks run slower near a black hole and space-time behaves like a glob of gelatin whose shape, or curvature, is determined by the presence of massive bodies: "space tells matter how to move; matter tells space how to curve." (J. A. Wheeler). The theory of General Relativity is, in essence, a mathematical formulation of this geometrical view of the Universe.

Ripples in Space-time

Einstein's model offers an elegant solution to the philosophical dilemma of Newton's "action at a distance". When a mass accelerates, its contribution to the underlying space-time also changes; like waves in a pond where a stone is thrown, this change propagates outward at the speed of light as ripples in the fabric of space-time: gravitational waves.

Everything produces gravitational waves: children playing a soccer game, cars on the highway, jet planes taking off and landing, volcanic explosions, they all induce vibrations in space-time, but these effects are extremely small. Only the dramatic acceleration of massive astrophysical objects could produce gravitational waves large enough to be detected on Earth. In this sense, the business of gravitational wave detection is poised to open new paths in the exploration of the universe. Our current knowledge of the universe is based on the observation of electromagnetic radiation, but gravitational waves are space itself changing its configuration, in response to the most violent rearrangements of mass and energy. Gravitational waves travel virtually undisturbed, unimpeded by intervening clouds of dust, and, if we learn how to listen, they can tell us tales of what happens in the core of a supernova or during the collision of two stars. Gravitational waves can teach us about the most mysterious astrophysical objects, such as black holes and neutron stars, and conceivably bring us information from the very beginning of the Universe.

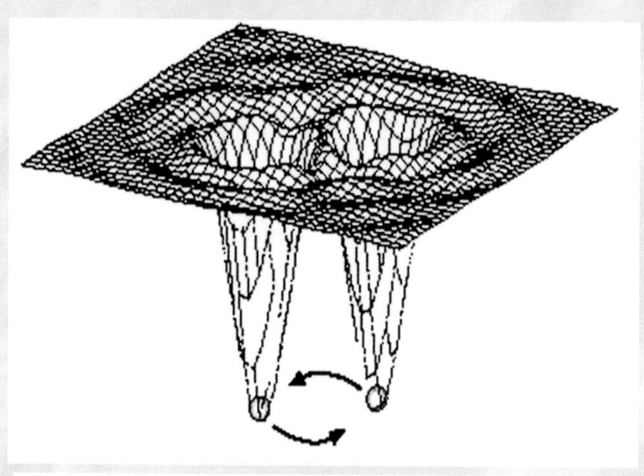

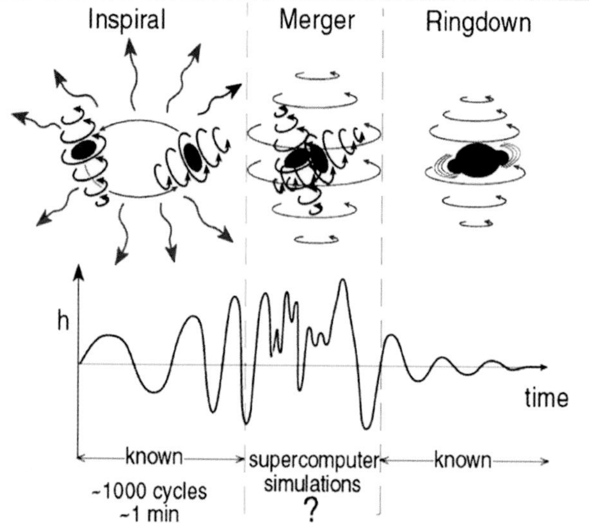

Top Right: *A binary system of compact massive objects, such as neutron stars or black holes, curves space-time. Additional ripples propagate outwards from the system, due to the rapid orbiting of the two objects. These perturbations are gravitational waves; their amplitude diminishes in inverse proportion to the distance from the source. Image courtesy Nick Strobel at www.astronomynotes.com*

Right: *Gravitational-wave emission from a compact binary system in the last few minutes of its life. This class of sources has a distinct signature and it is the one for which most predictions are formulated. The two massive object spiral faster and faster around each other, losing more and more energy to gravitational waves. The two bodies then collide and merge, leaving behind a black hole that vibrates to equilibrium. While "inspiral" and "ringdown" phases are well modeled, the merger phase is still being studied with supercomputer simulations. If the two objects are neutron stars, LIGO stands a good chance at detecting a GW emitted in the inspiral phase. If they are more massive black holes, the gravitational wave frequency is lower in inspiral phase and LIGO is more sensitive to what happens in merger and ringdown phase. Different analysis techniques have been implemented to identify these classes of signals, which are to date the most studied and best understood. According to current computations, LIGO stands a non-null chance to see one of these objects in its initial configuration and will see several within days in its advanced design. Image courtesy Kip Thorne.*

Indirect evidence of gravitational waves is already available, thanks to the observation of binary systems of neutron stars rotating around each other, where one of the neutron stars is a pulsar emitting regular pulses of radio waves, similar to the beacon of a lighthouse. The first such system to be observed was PSR 1913+16, whose discovery earned Russel Hulse and Joseph Taylor the 1993 Nobel Prize. Monitoring the radio pulses over 20 years, it was observed that the pulsar's 8-hour period of rotation sped up by 14 seconds, in beautiful agreement with General Relativity's prediction that the two stars are spiraling towards each other and shedding energy and angular momentum by emitting gravitational waves. The Hulse-Taylor pulsar tells us they exist, but gravitational waves have never been seen directly. As they travel through the Earth, they stretch and compress space in directions perpendicular to their propagation, locally altering the distance between free masses. This warping of space-time can be detected, if we can construct an instrument that can measure tiny changes in distance (strain) of the order of one part in 1,000,000,000,000,000,000,000.

A Laser Fishing Pole

The first attempt to directly measure gravitational waves dates to the 1960s, when Joe Weber, at the University of Maryland, connected a transducer to a 1.5-meter-long aluminum cylinder to monitor the changes in its length induced by the passage of gravitational waves. Through the years, a number of similar "resonant bar" detectors have been operated around the world, but none has yielded a confirmed detection of gravitational waves.

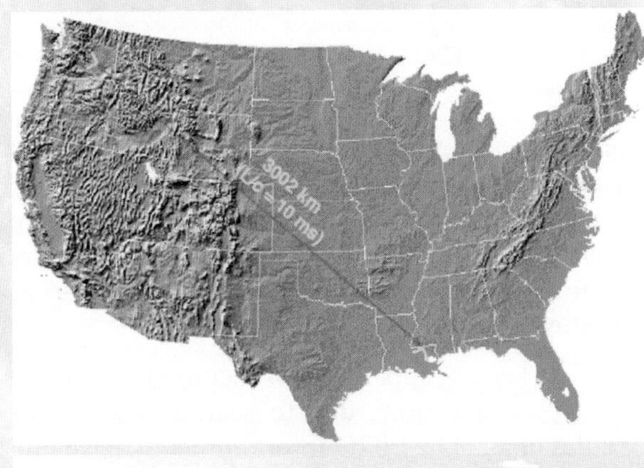

LIGO (the Laser Interferometer Gravitational-wave Observatory) is adopting a radically different approach. The LIGO project is sponsored by the National Science Foundation and jointly hosted by the California Institute of Technology and the Massachusetts Institute of Technology, in collaboration with a worldwide network of some 500 scientists (the LIGO Scientific Collaboration). LIGO uses laser light to measure changes as small as 10^{-18} m in the distance between mirrors that are suspended several kilometers apart. To achieve this sensitivity, one thousand times smaller than the size of a proton, LIGO scientist have worked at the frontiers of technology in optics, precision lasers, vacuum science and mechanical systems for more than 20 years.

The two LIGO observatories are located in Livingston, LA and in Hanford, WA Each LIGO detector is an interferometer, similar to the one used by Michelson and Morley in their 1887 ether experiment. The interferometer is laid out in an L-shape, with arms that are 4 km long, addressing the need for a long baseline to measure small strains.

A laser beam is split in two by a semi-transparent mirror, the beam splitter. The two components travel along the two perpendicular arms, are reflected from freely suspended mirrors at the far end, and are recombined when they return to the beam splitter. In normal conditions, the device is tuned so that the two recombining waves exactly cancel, with the peaks of one aligning with the valleys of the other in destructive interference. If the distance between the mirrors changes for any reason,

Left: *The location of the two LIGO sites in the USA and aerial photographs of the LIGO installations in Hanford, WA* ***(middle)*** *and Livingston, LA* ***(bottom)****.*

The Hanford facility comprises two co-located interferometers, sharing the same vacuum enclosure but otherwise independent, with 2 and 4 km baselines. Depending on its orientation, a gravitational wave traveling at the speed of light affects the two observatories with at most 10 ms delay, the time needed to cover a distance of about 3000 km. Images courtesy LIGO Laboratory.

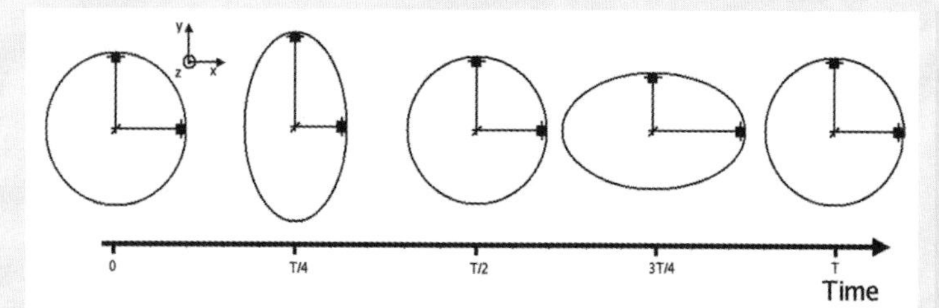

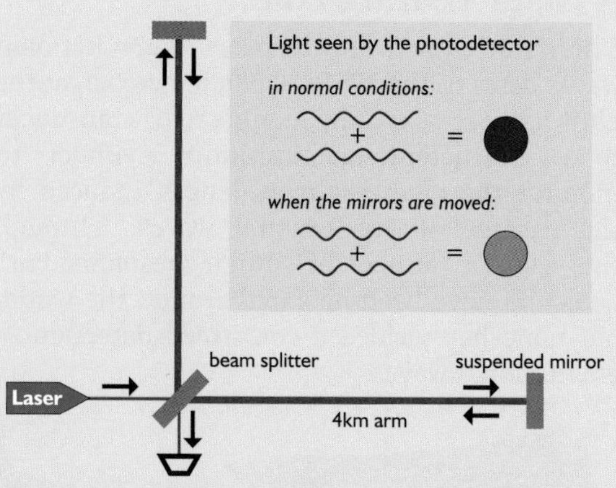

the two beams accumulate different phase delays. Peaks and valleys lose their alignment and just a bit of interference light leaks to be measured by a photodetector, signaling the passage of a gravitational wave. There are some additional design subtleties that increase the interferometer's sensitivity, such as bouncing the beams between the end and near mirror between 100 and 150 times in order to increase the length of travel and amplify the effect of a change in distance.

Fighting the Odds

An interferometer is essentially a complex machine that measures the distance between the beam splitter and two mirrors. There are several sources of noise competing with gravitational waves to perturb the position of the mirrors, but seismic vibrations are LIGO's principal nemesis.

The interferometers are built in isolated locations, in the woods of Louisiana and in the desert of Washington State, but they are so sensitive they can be affected by construction works, trains, trucks on the nearby highways, airplanes flying overhead, windstorms, ocean waves, and earthquakes around the globe. To isolate them from these low-frequency vibrations (below a few Hz), the mirrors are suspended by wires that are themselves cushioned by a system of masses and springs that absorb ground vibrations; the Louisiana facility is protected by an additional active insulation system, where hydraulic pressure is used to cancel the tremors coming from falling trees during logging in nearby forests.

Top Left: *Effect of the passage of a gravitational wave of period T traveling in a direction perpendicular to the plane of the paper: a circle stretches and compresses into an oval with the wave cycles, and free masses positioned on the circle get closer to and further from the center. In a laser interferometer such as LIGO, free masses are replaced by suspended mirrors and the arms change their length with the passage of the wave. The effect is extremely small: LIGO is trying to measure changes of the order of 1/1000 of the size of a proton over a 4-kilometer baseline. Image courtesy Laura Cadonati.*

Top Right: *Schematics of the LIGO interferometer. The laser beam is split in two components that travel along the perpendicular arms, are reflected by mirrors and recombine at the beam splitter. The interference pattern of the recombined beam is monitored by a photodetector. In normal conditions, the interferometer is tuned so that the two beams cancel each other and the photodetector does not see any light. When a gravitational wave modifies the length of the arms, the two beams accumulate phase changes and some intererence light is measured by the photodetector. Image courtesy Laura Cadonati.*

The laser beams travel in vacuum enclosures, to avoid distortions due to scattering on molecules of air. The LIGO interferometers have one of the world's largest ultra-high-vacuum systems, with a volume of nearly 300,000 cubic feet at one-trillionth of atmospheric pressure; evacuating and "baking out" the tunnels requires about six months. The mirrors are polished and coated for good reflectivity. Tiny magnets attached to the mirrors can be driven electromagnetically to adjust the mirrors' position and orientation. Dozens of control systems combine to hold all lasers and mirrors in proper alignment to within a fraction of a wavelength over the four kilometer lengths of both arms of the interferometers. The laser itself is state-of-the-art and the beam is stabilized to the long arm cavities of the interferometer to such an exquisite degree that, over one hundredth of a second, the frequency fluctuates by less than a few millionths of a cycle.

The result is a device with broadband high sensitivity: it can detect a strain of 10^{-21}m over a bandwidth of 100Hz. At low frequencies, below 10Hz, the sensitivity is limited by the seismic shaking of the mirrors. In the mid-range of frequencies, between 100 and 200 Hz, it is limited by thermal noise, the vibration of atoms in the suspension fibers. At larger frequencies, it is limited by quantum "shot" noise, as individual photons hit the photodetector and the mirrors like raindrops on a roof.

Finding gravitational waves in the midst of all this noise requires understanding both the device

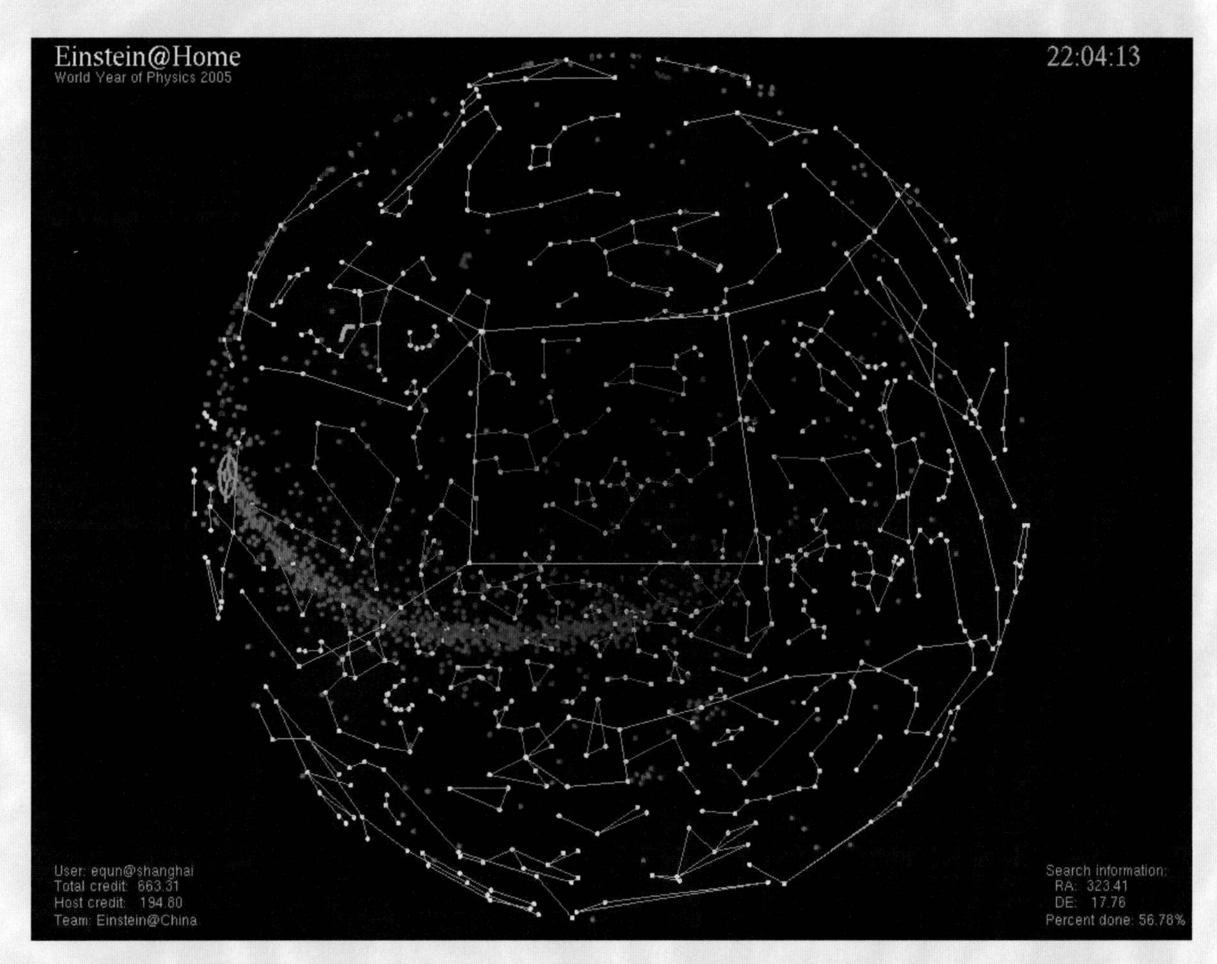

and the expected signal. The latter is not always possible: there is no good waveform prediction for most of the sources that LIGO is designed to discover. The ultimate defense for LIGO from the pollution of environmental and instrumental transients in the signal is the requirement of a coincident measurement in multiple detectors. The observatory comprises three interferometers, two of them co-located in Hanford (one of them with a shorter, 2km baseline) and the other in Louisiana. A gravitational wave will appear in both sites within 10 milliseconds (the time it takes a gravitational wave to travel from one to the other site at the speed of light) and with the same calibrated amplitude and at the same time in the two Hanford interferometers.

From Development to Exploration

In the fall of 2005, the LIGO detectors reached their sensitivity design goals and began long-term data acquisition, listening for gravitational waves produced by sources at distances of tens of million light-years. Astrophysicists, hardware experts and data analysts are sifting through the daily generated terabyte of LIGO data, digging in the noise to identify gravitational wave signatures. The data they use includes information on the distance between mirrors (the "gravitational wave channel"), sampled at 16,384 Hertz, and signals from hundreds of data acquisition channels for diagnostics of the machine and monitoring of the environmental noise.

LIGO is listening to space, searching for the chirp wave produced by the death spiral of binary neutron stars before they collide and merge into a black hole, for the sound of an exploding star in a supernova, for the first cries of nascent neutron stars and their instabilities, for the periodic humming of pulsars, for the murmurs of the relic gravitational radiation from the Big Bang and the primordial universe, and for all serendipitous sources that may be out there. LIGO scientists, like early explorers, are open to the unexpected, probing the universe in ways that were not done before.

■ ***Above:*** *Screensaver for the einsten@home project. The anlysis of LIGO data is a monumental task. Computer users from around the world can now contributed the unused CPU time on their desktops and laptops to the search for gravitational waves from pulsars and supernova remnants. For more information, visit* ***www.einsteinathome.org.*** *Image courtesy Bruce Allen, David Hammer and Eric Myers.*

The First Detection

When will LIGO find its first gravitational wave, and what will its source be? There are too many uncertainties in current astrophysical models to give a good answer to this question – we do not know how many black-hole or neutron-star binary systems are in the nearby galaxies, what happens during a stellar core collapse or whether spinning neutron stars have blemishes on their surfaces. Any asymmetry in the evolution of these systems causes dramatic changes in the pattern and amplitude of gravitational wave emission. Super-computers are hard at work to provide answers and models, using numerical General Relativity, but there are no firm answers yet. Since no one has ever listened for signal this faint, we cannot know beforehand what we will hear.

LIGO may not yield detection for a few more years, but an upgraded LIGO should hear waves from in-spiraling binary neutron stars within days. Advanced LIGO will be ten times more sensitive and will listen to sources in a volume of space a thousand times larger, the installation of a new laser, new optics suspensions and isolation systems that will allow this enhanced sensitivity should start within the next five years.

And LIGO is not alone in this quest: other interferometers are operating or about to start in Europe (GEO and VIRGO) and in Japan (TAMA). Protocols for data exchange and joint analysis are being tested on short duration runs, proving this is truly a worldwide cooperative effort.

The LISA experiment will launch a laser interferometer into space – the ultimate cure for seismic noise. A growing community of scientists is now committed to the quest for gravitational waves: the first discovery will only be the beginning for a new branch of science that may take us beyond General Relativity and unveil a whole new universe.

For further information, consult these websites:

LIGO web site:
http://www.ligo.caltech.edu/

LIGO Scientific Collaboration web site:
http://www.ligo.org/

To participate in LIGO data analysis:
http://www.ligo.org/

Exhibit on Gravitational Waves at the American Museum of Natural History:
http://sciencebulletins.amnh.org/astro/f/gravity.20041101

Above: *LISA is as ESA-NASA mission involving three spacecraft flying approximately 5 million kilometres apart in an equilateral triangle formation. Together, they act as a Michelson interferometer to measure the distortion of space caused by passing gravitational waves. Lasers in each spacecraft will be used to measure minute changes in the separation distances of free-floating masses within each spacecraft. Image courtesey ESA.*

■ *This visualization shows what Einstein envisioned. Using the Columbia supercomputer at the NASA Ames Research Center, researchers created a three-dimensional simulation of merging black holes. Previous simulations had been plagued by computer crashes, but scientists at NASA's Goddard Space Flight Center found a method to translate Einstein's equations in a way that computers can understand. According to Einstein's theory, when two massive black holes merge, the surrounding space shakes to and fro like a bowl of Jell-O as gravitational waves race out from the collision at light speed. Image courtesy Henze, NASA.*

9

Detecting HEAT RADIATION from Space...

Check out what's cool with the infrared Spitzer Space Telescope this year. *Michelle Thaller*, an expert at conveying information to the public in an exciting way, as well as being an astronomer, reviews the amazing discoveries coming from Spitzer. The images are dramatic. The science is astounding.

...The SPITZER Space Telescope

NASA's Great Observatory Program had an ambitious goal: launch and operate large space telescopes that would allow astronomers to view the universe across the entire electromagnetic spectrum, from gamma rays to the infrared. Whenever possible, the observatories would take complementary data, mapping the galactic plane together or simultaneously viewing a transitory event such as a comet or supernova explosion. In the end, the observatories would create the richest database ever made available to astronomers, allowing them to piece together a more complete view of the universe, using clues taken from one end of the spectrum to the other

Past Great Observatories (by order of wavelength) included the Compton Gamma Ray Observatory, the Chandra X-Ray Observatory, and the Hubble Space Telescope.

The Spitzer Space Telescope was the final jewel in the Great Observatory crown, designed to view the universe in light ranging in wavelengths from 3 to 160 microns, the entire span of the infrared spectrum. Infrared light covers a range of wavelengths 30 times greater than visible light, and the infrared universe hold its own particular mysteries. At shorter wavelengths, infrared light is able to pass through obscuring dust and gas, allowing astronomers to probe the interior of star-forming clouds or view the hidden core of the Milky Way galaxy. At longer wavelengths, lower temperature material begins to glow in its own right, revealing cool dust and gas structures in interstellar space, as well as planet-forming disks around young stars. Launched on August 25th, 2003, Spitzer has been continually churning out amazing and surprising discoveries. It has turned its sensitive infrared instruments to celestial objects ranging from nearby comets in our own Solar System to the most distant galaxies yet imaged, and changed our view of these and everything else in between.

For a physically small spacecraft, Spitzer packs a lot of punch. Spitzer boasts a particularly innovative design, which allows the telescope to passively cool itself to very low temperatures without needing heavy, and therefore expensive, thermal shielding. The spacecraft has only one heat shield, which also serves as Spitzer's energy source, using solar arrays on the opposite side. The observatory is kept cool by keeping the heat shield continuously between the Sun and the thermally delicate infrared telescope assembly. As one side of the telescope is kept continually in shadow, it can be used to dissipate excess heat from the system. The telescope's two-

Left: *The Spitzer Space Telescope under construction. Image courtesy Russ Underwood, Lockheed Martin Space Systems.*

■ *Artist's concept of Spitzer's Earth-trailing orbit. Image courtesy NASA/JPL-Caltech.*

tone color scheme allows it to passively cool to fifteen degrees Kelvin (or fifteen degrees above absolute zero) just by radiating excess heat away through the black paint-covered side of the spacecraft. The reflective paint on Spitzer's other side allows any stray light from the Sun to bounce harmlessly off the spacecraft, further protecting the sensitive instrumentation.

Spitzer's 86 cm-diameter beryllium primary mirror is another engineering triumph, designed to distort into near-perfect curvature when the telescope reached 10 degrees Kelvin. Below the mirror is a chamber containing three of the most sensitive infrared instruments ever built: the Infrared Array Camera (IRAC), the Infrared Spectrograph (IRS), and the Multi-band Imaging Photometer for Spitzer (MIPS). IRAC is a four-channel camera that provides simultaneous images at 3.6, 4.5, 5.8, and 8 microns. The IRS consists of four separate spectrographs, covering the range of 5.3 to 40 microns, and MIPS fills out the longer wavelength space, producing images and photometry at 24, 70, and 160 microns.

Each of the instruments is cooled by a liquid helium cryostat which allows internal temperatures to drop to a few degrees Kelvin. As the telescope has so little external shielding, it was necessary to get the observatory as far away from the warm environment of the Earth and Moon as possible. To do this, Spitzer was launched into an Earth-trailing solar orbit, lagging behind the Earth as it spins around the Sun. In this orbit, Spitzer continuously drifts farther and farther away from the Earth, and is now many times farther from Earth than the Moon. Obviously, in this sort of orbit, any kind of servicing mission is out of the question. Eventually, the liquid helium cryogen will run out and Spitzer's instruments will begin to warm. Current estimates give Spitzer a primary lifetime of over five years, and plans are now being laid for an extended mission. Even without the cryogen, Spitzer has the ability to keep cool through passive cooling. This would enable the IRAC camera to operate for several more years, providing astronomers with the best infrared observations available until the next generation of infrared observatories is launched.

Witness to Star Birth

One of Spitzer's main science goals was to map star formation inside the vast interstellar molecular clouds of our galaxy. Indeed, one of the first images taken by Spitzer was of a cloud called IC 1396. Spitzer's image of this star-forming region shows the power of combining several instruments to create a revealing composite image from several different wavelengths of light. The smaller inset images show the cloud as observed by two instruments, IPAC and MIPS. In the IRAC image, shorter wavelength infrared light allows us to see the warm knots of star formation inside the obscuring dust of the cloud. At longer wavelengths, MIPS picks up emission from the dark, cold material of the cloud itself. Combining the images allows us to see about a dozen newly-discovered protostars, the hydrogen gas contained in the cloud, and thermal emission from organic molecules in the dust, in this case, polycyclic aromatic hydrocarbons. Using IRS, astronomers were able to probe the chemical environment inside similar clouds, finding compounds ranging from water ice, to methyl alcohol, to acetylene and hydrogen cyanide, which are precursor molecules to the formation of DNA.

Subsequent observations of star-forming clouds provided new views of old astronomical favorites. For example, false-color Spitzer images reveal a different side of the Trifid Nebula. In visible light, dark lanes of dust can be seen obscuring parts of the glowing nebula. In the infrared, the dark lanes become bright, revealing the inner structure of the cloud. All together, Spitzer uncovered 30 massive embryonic stars and 120 smaller newborn stars throughout the Trifid Nebula, in both its dark lanes and luminous clouds.

In some cases, the sheer beauty of these clouds took everyone's breath away. In this image of a cloud named NGC 1333, a flurry of star formation sparkles before our eyes. This cloud is almost entirely obscured by dust in visible light, but infrared reveals dozens of new stars, many with jets from their poles, scattered throughout the interior of the cloud.

Planets in all the Wrong (and Right) Places

One of the fundamental questions of modern astronomy is how many stars have planets orbiting around them. In visible light, extrasolar planets are nearly impossible to detect directly, as the weak visible light they reflect is far too

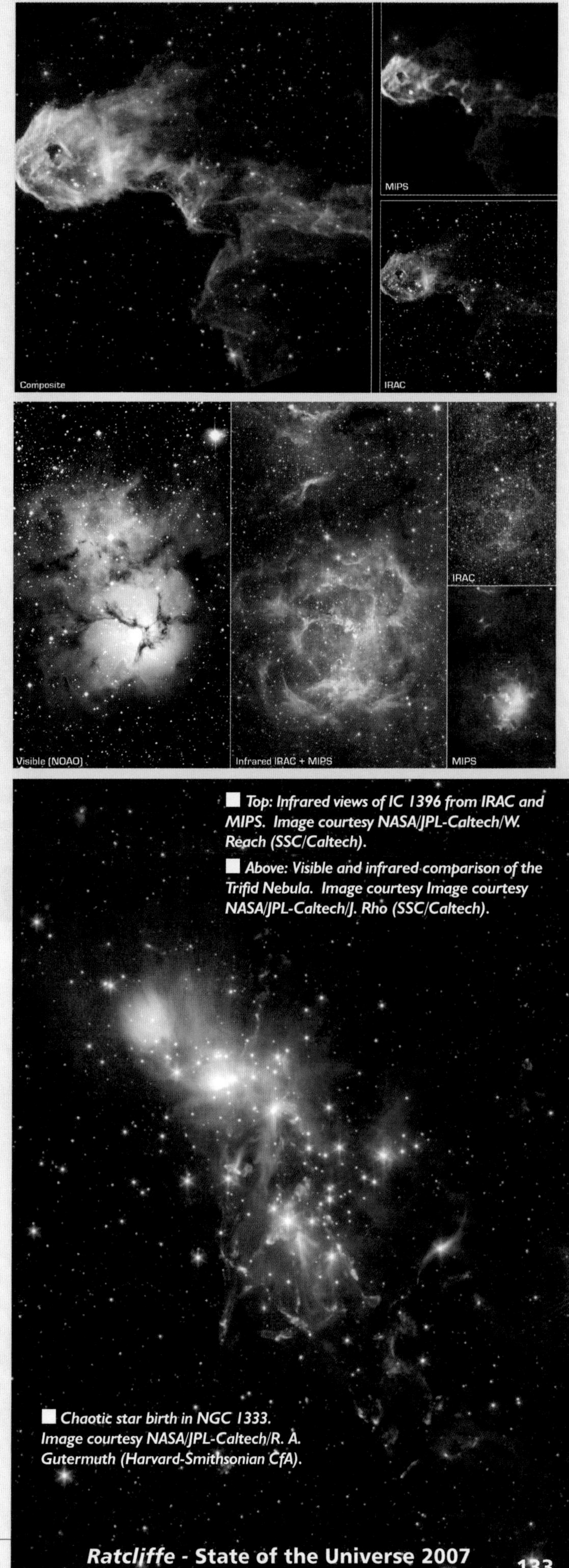

Top: Infrared views of IC 1396 from IRAC and MIPS. Image courtesy NASA/JPL-Caltech/W. Reach (SSC/Caltech).

Above: Visible and infrared comparison of the Trifid Nebula. Image courtesy Image courtesy NASA/JPL-Caltech/J. Rho (SSC/Caltech).

Chaotic star birth in NGC 1333. Image courtesy NASA/JPL-Caltech/R. A. Gutermuth (Harvard-Smithsonian CfA).

■ *Artist's concept of young planets forming a gap in a circumstellar dust disk. Image courtesy NASA/JPL-Caltech/T. Pyle (SSC/Caltech).*

■ *Inset: Artist's concept of a planetary disk around a brown dwarf. Image courtesy NASA/JPL-Caltech/T. Pyle (SSC).*

dim to detect against the glare of their stars. In the infrared, however, there are two methods for finding planets. In the first case, Spitzer is able to detect infrared emission from warm dust orbiting stars. These dusty disks break down into two distinct groups: protoplanetary disks and debris disks.

A protoplanetary disk, as its name implies, is still in the planet-forming phase of development. Dust and gas left over from the formation of the central star settles into stable orbits around the star, and over time, begin to form clumps under the influence of gravity or density waves in the disk. As planets form, they hollow out regions of the disk, forming gaps in the dust and gas near them, similar to gaps in the rings of Saturn which are caused by its moons. Spitzer is able to detect these gaps and gauge where active planet formation is taking place. In one case, Spitzer observed a protoplanetary disk around the young star CoKu Tau 4. Although the star is thought to be only a few million years old, regions of its disk have already been cleared, presumably by planet formation. This result caused quite a stir, as previous models of stellar disks did not predict large-scale planet formation at such an early age.

More surprises were in store when Spitzer began to observe planet-forming disks around unlikely parent stars. Many astronomers had assumed that only sun-like stars would have protoplanetary disks. If a star were much less massive the Sun, it might not have the gravity to attract a dust disk of any substance. If it were more massive, the extreme stellar winds and radiation pressure would surely blow the dust back into interstellar space. In the end, both scenarios appear to be wrong. In one case, Spitzer found a healthy protoplanetary disk around an object that barely qualifies as a star. Cha 110913-773444 is a brown dwarf, or a star that never had sufficient mass to ignite hydrogen fusion in its core. At only about eight times the mass of Jupiter, this object is substantially less massive than even other extrasolar planets previously detected. But amazingly, it seems to be forming its own planetary system. On the other side of the spectrum, Spitzer is also finding planetary disks around the largest and brightest stars in the universe. Two hypergiant stars in the Large Magellanic Cloud, R66 and R126, both appear to have bloated disks that begin hundreds of astronomical units away from their stars, just outside the range of scorching, dust-destroying, radiation.

The other class of circumstellar disks that Spitzer routinely observes are debris disks, thin disks of dust that are either the remnants of past planet formation, or the debris created by interactions between mature planets. In the case of Vega, Spitzer imaged a huge dust ring extending 815 astronomical units from the central star. This disk is unstable, and it being blown away by Vega's intense heat and stellar winds. Astronomers think that the disk is the leftover debris from a dramatic collision, caused by two objects the size of Pluto colliding and nearly obliterating one another only a million or so years ago.

In some cases, astronomers were surprised by how similar debris disks in other planetary systems were to our own Solar System, and also how different. In the nearby system of HD69830, a star of similar mass and age to our Sun, Spitzer found evidence of an "asteroid belt" very much like our own. Our asteroid belt is thought to have been formed by a disrupted protoplanet – a small object that got caught in a tug-of-war between opposing gravities of the Sun and Jupiter. In this system, however, the debris belt is 25 times more massive than our asteroid belt, and light scattered through its disk would dominate the night sky. This gave astronomers pause to imagine what an alien sunset might be like on a world in that system.

But the biggest surprise turned out to be that Spitzer was sensitive enough to detect infrared light directly from extrasolar planets – literally the first photons detected not from a star, but a planet in another system. So far, all the extrasolar planets we know of have been discovered indirectly, mainly by watching the central star wobble under the gravitational influence of unseen planets. Astronomers already knew that three star systems, HD 209458, TrES-1, and HD 189733 had large, "hot Jupiter"-type planets, gas giants that zip closely around their parent stars. In these toasty orbits, the planets soak up warmth from their parent star and shine brightly in the infrared.

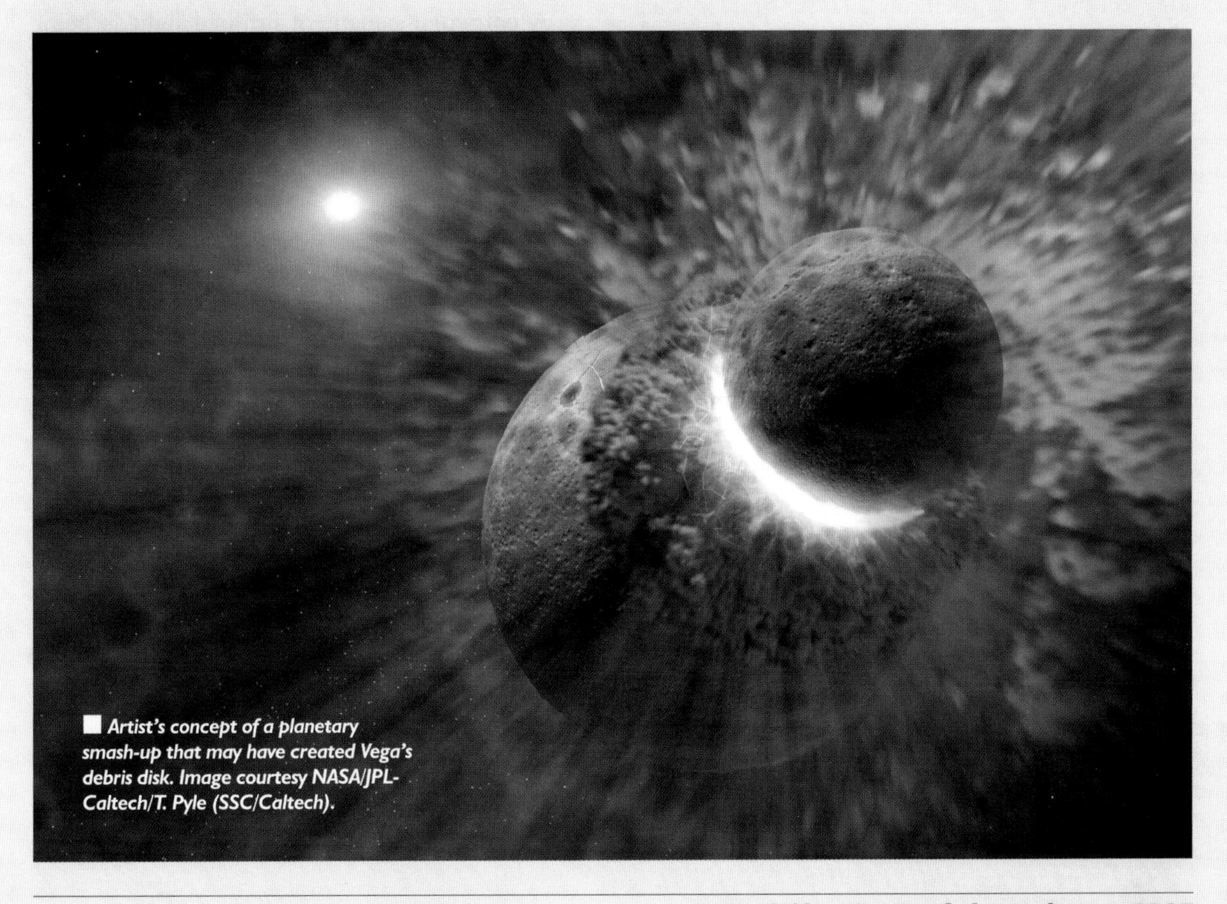

Artist's concept of a planetary smash-up that may have created Vega's debris disk. Image courtesy NASA/JPL-Caltech/T. Pyle (SSC/Caltech).

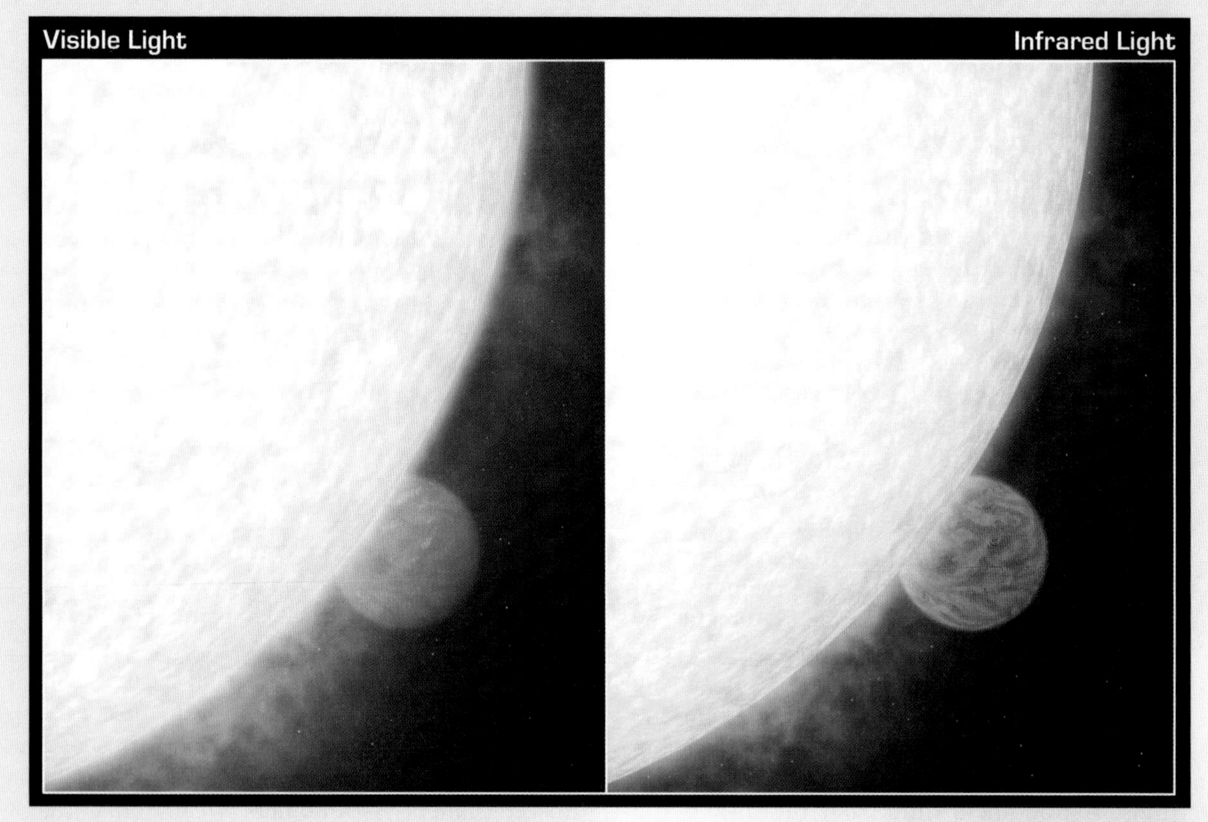

But what was important about these three systems was a simple matter of luck: all three planets, when viewed from our solar system, disappear behind their parent stars during their orbits. In other words, each time the planets make one orbit, they move behind and are "eclipsed" by their own stars. Spitzer was able to measure a drop in the total amount of infrared light coming from each system during such an eclipse. To distinguish between infrared light coming from the planets as opposed to the fiery hot stars, the astronomers used a simple trick. First, they used Spitzer to collect the total infrared light from both the stars and planets. Then, when the planets dipped behind the stars Spitzer measured the infrared light coming from just the stars. This pinpointed exactly how much infrared light belonged to the planets. The Spitzer data indicates that all these planets are at least a steaming 700 degrees Celsius – hot indeed!

The Internal Structure of Galaxies

For decades, astronomers have classified galaxies according to their shapes, and have sought clues to whether there was an evolutionary sequence present. For example, are large elliptical galaxies formed when several smaller galaxies merge together? Are spiral arms also formed when two or more galaxies interact, creating density waves in the galactic dust and gas? Spitzer has provided important clues to these and other galactic mysteries, but as so often happens in science, the new observations are also making the whole picture a lot more complicated.

A particular benefit of infrared light is being able to see the underlying structure of dust and gas in a galaxy, not just the light coming from existent stars. In a way, this allows astronomers to glimpse the "skeletal" structure of the galaxy and see not just where stars are forming, but where the majority of the galactic substance is underneath.

A Spitzer observation of the spiral galaxy M81 is a good example. A visible-light image of Messier 81 is shown in the upper right inset on the next page. Both the visible-light picture and the 3.6-micron near-infrared image (colored blue) trace the distribution of stars, although the Spitzer image is virtually unaffected by obscuring dust. Both images reveal a very smooth stellar mass distribution, with the spiral arms relatively subdued.

As one moves to longer wavelengths, the spiral arms become the dominant feature of the galaxy. The 8-micron emission (green) is dominated by infrared light radiated by dust that has been heated by nearby stars. The dust particles are composed of silicates (chemically similar to beach sand) and organic molecules and provides a reservoir of raw materials for future star formation. The 24-micron

Above: *Comparison of an extrasolar planet observed in visible and infrared light. Image courtesy NASA/JPL-Caltech/R. Hurt (SSC).*

image (orange) shows emission from warm dust heated by the most luminous young stars. The infrared-bright clumpy knots within the spiral arms show where massive stars are being born in giant H II (ionized hydrogen) regions. Studying the locations of these star-forming regions with respect to the overall mass distribution and other constituents of the galaxy will help astronomers identify the conditions needed for star formation.

In another example, two of the Great Observatories pooled their resources to create a mind-blowing image of one of the best-known galaxies in the sky. Spitzer and Hubble joined forces to create a striking composite image of Messier 104, commonly known as the Sombrero Galaxy. In Hubble's visible light image, only the near rim of dust can be clearly seen in silhouette. Spitzer's image revealed a bright, smooth ring of dust circling the galaxy, seen in red. Spitzer was also able to peer through the dusty plane of the galaxy, revealing an otherwise hidden disk of stars within the dust ring. Like many spiral galaxies, the Spitzer data showed that the Sombrero Galaxy's disk is warped, which is often the result of an encounter with another galaxy.

But again, just like Spitzer's observations of planets, bigger surprises were yet to come. Everything came to a head when Spitzer observed a famously active (and mysterious) galaxy called Centaurus A. In visible light, Centaurus A appears to be a large, blob-shaped, elliptical galaxy with a dramatic dust lane spanning its center. Astronomers had several clues that Centaurus A had been involved in galactic collisions in the past. For one thing, its weird morphology suggested that something had disrupted the galaxy's original shape. Another clue is that Centaurus A is pouring out all kinds of radiation from its obscured core, suggesting that a giant black hole is lurking inside and being well-fed by material collected in a past galactic merger. But no one expected the dramatically well-organized interior that Spitzer revealed its infrared image. Underneath the thick dust lane, astronomers imaged an unusual parallelogram-shaped structure. The shape can be well-explained by using a model of a flat spiral galaxy falling into an elliptical galaxy and becoming twisted and warped in the process. The folds in the warped disk, when viewed nearly edge-on, take on the appearance of a parallelogram. This model predicts that the leftover galaxy will ultimately flatten into a plane before being entirely devoured by Centaurus A.

Top: *Visible and infrared comparisons of M81. Image courtesy NASA/JPL-Caltech/K. Gordon (University of Arizona) & S. Willner (Harvard-Smithsonian Center for Astrophysics), N.A. Sharp (NOAO/AURA/NSF).*

Middle: *Visible and infrared images of the Sombrero Galaxy, M104. Image courtesy NASA/JPL-Caltech/R. Kennicutt (University of Arizona), Hubble Space Telescope/Hubble Heritage Team.*

Bottom: *The unusual parallelogram feature in the Spitzer image of Centaurus A. Image courtesy NASA/JPL-Caltech/J. Keene (SSC/Caltech).*

And the list goes on. Spitzer has also been instrumental in determining that the Milky Way galaxy is indeed a barred spiral galaxy, something that has long been suspected. In an artist's conception based on Spitzer data, we can see our true place in the galaxy, as well as the orientation of the central bar of stars. Other amusing discoveries include a spiral galaxy with only one spiral arm. In the end, Spitzer may challenge the very notion of galaxy classification, as some galaxies appear to have one shape when viewed in visible light, and another underlying structure when seen in the infrared.

The Distant Universe

After the stunning success of the Hubble Deep Field, the practice of using large amounts of telescope to time to peer into the most distant recesses of the universe has become something of a cottage industry. Here again, the Great Observatories came together to combine their unique views to give us a more complete picture of the early universe. In an observing program called "the Great Observatory Origins Deep Survey" the three functioning Great Observatories, as well as the European Space Agency's XMM-Newton mission, spent literally weeks staring at the same small, relatively empty, patch of the sky. The goal was to push each observatory's sensitivity to the limit and make the deepest observation of the universe yet made.

Here again, the complementary differences between the Great Observatories led to the biggest surprises. For some time, astronomers had been perplexed by very distant galaxies that were detected using Chandra, but were invisible to Hubble. X-rays, detectable to Chandra, are produced by extremely hot, million-degree gas. What could produce such a strong flux of X-rays, yet still be completely dark to Hubble? In distant active galactic nuclei, X-ray emission is thought to come from hot regions in an accretion disk around a massive black hole. But what was blocking all the visible light that must also be generated from glowing gas in the disk? Perhaps the cores of these young galaxies, seen as they were almost ten billion years ago, are completely obscured by thick dust clouds. If so, the hot dust should radiate plenty of infrared light. Spitzer was able to detect the X-ray bright, but optically invisible galaxies, indicating that such a scenario was likely the case. In the early universe, galaxies seem to be less well-organized, with plenty of material falling down the central black holes. The active cores must be hidden behind immense dust clouds, which completely absorb the visible light generated by the accretion disks.

Infrared holds several specific advantages over visible light surveys of the deep universe. For one thing, distant galaxies are heavily red-shifted, meaning that photons originally generated in the galaxies as visible or even ultraviolet light have been stretched out by their journey through the expanding space of the universe. When the photons are picked up by our telescopes, billions of light-years away, they have lost so much energy that they are now detectable in the infrared. In a real sense, infrared telescopes can see farther out into the universe, and therefore farther back in time, than visible instruments. And of course, Spitzer's observations changed our picture of the first billion years after the formation of the universe. For one thing, galaxies seem to have formed much more quickly than previously thought.

One of the most distant galaxies imaged by Spitzer, HUDF-JD2, seems to have accumulated an amazing amount of material in a very short time. Astronomers nicknamed the galaxy the "big baby." Somehow, the galaxy was able produce eight times more stars than our own Milky Way galaxy within the first few hundred million years after the big bang. The galaxy is believed to be as far away as the most distant known galaxies. It represents an era when the universe was only 800 million years old, about five percent of the universe's age of 14 billion years. And again, this calls into question our current theories of galaxy formation. So many distant, therefore young, galaxies seem to be involved in mergers that scientists are beginning to suspect that larger galaxies are built up over time by the conglomeration of many smaller galaxies. But the big baby galaxy seems to suggest that some of the first galaxies in the universe became surprisingly massive right away. Either way, astronomers now have a better, if more complicated idea about how the first galaxies formed over 13 billion years ago.

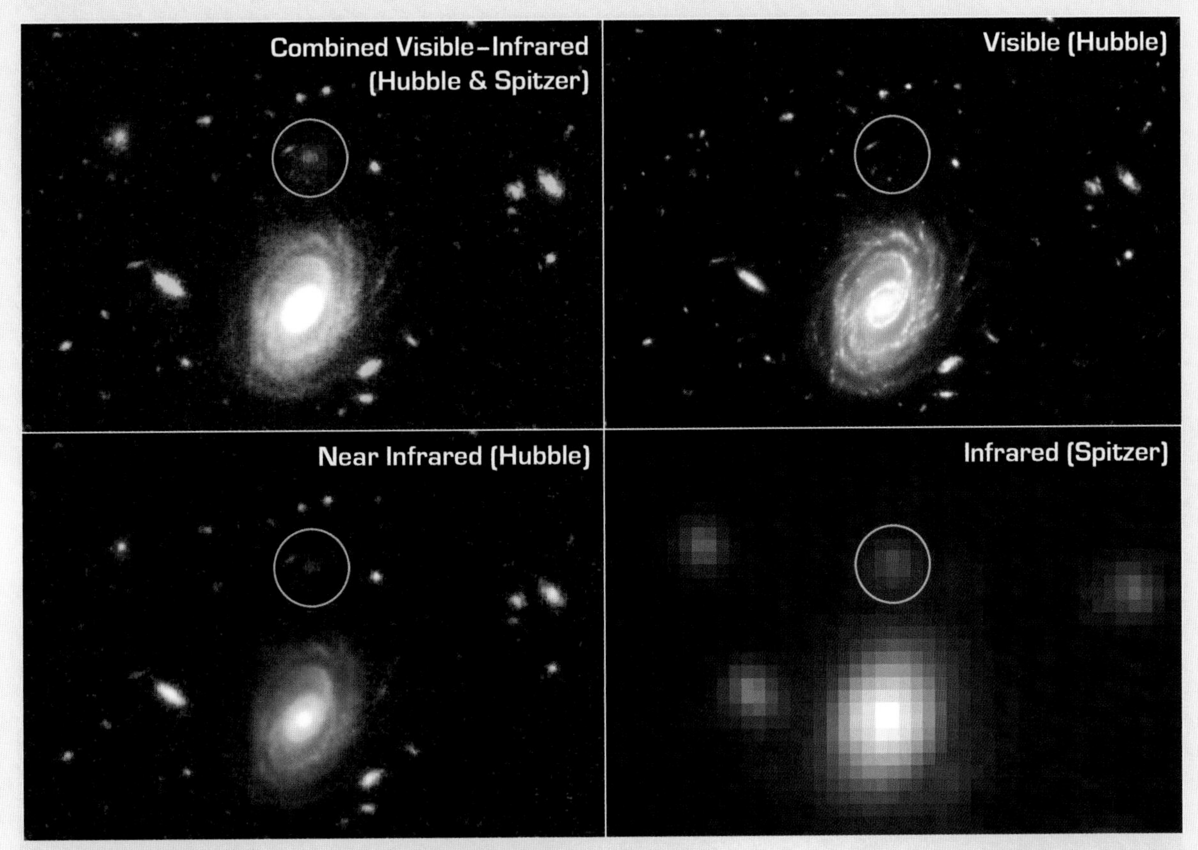

At the very edge of the universe, Spitzer may also have made an incredible discovery by detecting infrared radiation so distant and faint, it's likely to be from first stars ever to form. Behind all of the bright nebulae and faint galaxies, Spitzer is detecting a dim glow coming from all areas of the sky that can not be attributed to known stars and galaxies. The objects producing this "infrared background" radiation are so small and faint, Spitzer cannot resolve them into individual objects. And although Spitzer will likely spend some of its next years of operation confirming this, astronomers believe they might be glimpsing light generated in an era 13 billion years ago when, after the fading embers of the Big Bang gave way to millions of years of pervasive darkness, the universe first came alive. These first-generation stars, called Population III stars, were probably hundreds of times more massive the Sun and lived violent, brief, lives before exploding into supernovae.

If the ultraviolet light Population III stars emitted over 13 billion years ago could be detected, it would be red-shifted into the infrared. This infrared background was hinted at by data taken by NASA's Cosmic Background Explorer (COBE) satellite in the 1990s, but the observation needed follow-up. The low noise and high resolution of Spitzer's infrared array camera enabled astronomers to remove the fog of foreground stars and galaxies, revealing the dim glow beneath. Now, astronomers feel they are hot on the trail of the very first stars that ever glowed. Spitzer will not be able to resolve this background, but future infrared missions, like NASA's James Webb Space Telescope, will likely be able to image clusters of these stars, and perhaps even witness some of the supernovae that produced the universe's first black holes.

By any definition, the Spitzer Space Telescope has been a tremendous success. Many of its discoveries were anticipated, like the spectacular views Spitzer has given us of the interior of star-forming clouds. Other results struck astronomers completely by surprise; not even Spitzer's designers had thought that their telescope would be the first to detect direct light from an extrasolar planet. So now, with a few more years of Spitzer's primary mission to plan, astronomers are looking toward the future. Two great new infrared observatories are currently under development: NASA's James Webb Space Telescope and the European Space Agency's Herschel/Planck Telescope. The promise of these two missions is astounding; we may literally be able to see back to the time when the first stars in the universe began to glow. Spitzer has revealed the tremendous potential of infrared observatories, and the best is yet to come.

■ ***Opposite*** *Artist's concept of the Milky Way galaxy, showing the central bar and our location in the spiral arms. Image courtesy NASA/JPL-Caltech/R. Hurt (SSC).*

■ ***Above*** *: Visible and Infrared views of the 'big baby' galaxy. Image courtesy NASA, ESA/JPL-Caltech/B. Mobasher (STScI/ESA)*

10
Those STRANGE NEWS STORIES out there...

For every bit of good astronomy news we get, there is always some bad astronomy lurking in the shadows, and this past year was no exception, as *Phil Plait* explains.

...a year of BAD ASTRONOMY

EVERY YEAR brings new advances in astronomy. Better telescopes, better detectors, new and clever ideas... they all contribute to our understanding of the Universe.

But like anything else humans do, two steps forward always seem to be followed by one step back: for every bit of good astronomy news we get, there is always some bad astronomy lurking in the shadows, and this past year was no exception. If anything, some of the spinning, folding, and mutilating of astronomy was on a much larger scale than in previous years. Sure, television, Hollywood, and the news media did their share of mangling good science, but now even the government has joined in the battle. Scientists have had their hands full battling this nonsense on all fronts. What follows here is simply a taste of the silliness we've had to face over the past 12 months.

Bad Astronomy in the News

Nothing gets people more excited than the idea that at any minute, the Earth could be destroyed by some giant cloud of energetic plasma spat out from the galactic core. Well, OK, maybe that's a bit of an exaggeration. But certainly it's no more overblown than the breathless article carried by the Yahoo!News website entitled:

"PLANET-DISSOLVING DUST CLOUD IS HEADED TOWARD EARTH!"

(yes, it was in bold all-cap letters)

[link: http://tv.yahoo.com/news/wwn/20050912/11265372000 1.html].

The article starts off with a bang:

"Scared-stiff astronomers have detected a mysterious mass they've dubbed a 'chaos cloud' that dissolves everything in its path, including comets, asteroids, planets and entire stars – and it's headed directly toward Earth!"

It goes on from there, describing how it was discovered, how it fits in with many cutting-edge theories, and, sadly, what the implications are: "The bad news is that the total annihilation of our solar system is imminent."

The only problem with this is that it's complete fantasy.

First, there's no such thing as a "chaos cloud". The article describes it as being a seething mass of particles created near a black hole in a process called Hawking radiation. While the process itself is theoretically possible – it's a marriage of Einstein's equations of relativity with the quantum mechanical laws of microscopically small scales —it's never been observed, and it's incredibly slow. It would take billions of years (longer than the lifetime of the Universe) to create more than a handful of subatomic particles this way.

The news article gets worse from there. The only way to save ourselves, it says, is to create a sort of "space ark", which can carry all the inhabitants of Earth safely away. The destination recommended? The Andromeda Galaxy.

The problem here is... well, everything. First, we have no way to build a space ark at all. Then, even if we could, how big would one have to be to

A 'chaos cloud' approaches . . .

carry 6.5 billion people? Third, the Andromeda Galaxy is about 2.9 million light years – a mind-numbing 17 million trillion miles – away. Getting an ark to go that far is a little bit ahead of our current technology. Like, millennia ahead.

The article is full of nonsense like that, but it's reported straight, as if it's real. What news source could possibly have an article like this printed seriously?

If the first thing you thought of was "The Weekly World News", then give yourself a gold star. This is the same bastion of supermarket checkout lanes that brings you BatBoy (a human/bat hybrid) and aliens shaking hands with Presidents on a weekly basis. If the WWN ever had a story in it that wasn't totally made up, that would be news.

People reading the WWN either understand it's supposed to be goofy, or else they take the WWN seriously and deserve what they get. But Yahoo!News picked the piece up and ran it straight. Shame on them.

Still, people who read it and took it seriously – and a lot of people did – should really have taken a moment to think about what they were reading, and maybe done a quick web search on some of the "facts" in the article. After all, if the whole world is at stake, a little googling doesn't seem like such a hardship, now does it?

Bad Astronomy Via Email

In August, 2005, an email about Mars was spread around the internet. The subject line read, **"The Red Planet is about to be spectacular!"**, and it described how Mars would be closest to Earth on August 27. It detailed how to view it, and why it would be so close. But it also very strongly implied that Mars would be huge in the sky, as big as the full Moon!

Of course, the email was wrong.

Mars and Earth both orbit the Sun, so, like two cars on a racetrack moving at different speeds, sometimes they are close together, and sometimes farther apart. Every year and a half or so Earth catches up to Mars, and Mars will appear bigger through a telescope.

Note the key phrase there, "through a telescope". Mars is pretty dinky as planets go, about half the Earth's diameter. Even at its closest, it's tens of millions of miles away. That means that you need a pretty good telescope to see any features on Mars at all. To the naked eye, Mars is never more than a red dot in the sky.

The reason people thought it would be big was because of this line in the email:

The encounter will culminate on August 27th when Mars comes to within 34,649,589 miles of Earth and will be (next to the moon) the brightest object in the night sky. It will attain a magnitude of -2.9 and will appear 25.11 arc seconds wide. At a modest 75-power magnification

Mars will look as large as the full moon to the naked eye.

Note the paragraph break before that last line. If you just skim email (as most people do), you might miss the very important sentence that says that through a telescope at 75x power, Mars will look as big as the Moon does to your naked eye. It's very easy to misread that and think that to the naked eye, Mars will look as big as the Moon does. And so an internet legend was born…

… or reborn, more accurately. Because not only is this email wrong, it's not even original. It's a copy of an email that was sent out back in 2003, the last time Mars approached Earth in its game of cosmic race cars. And the email wasn't accurate even then! The numbers it gave were incorrect, and again it strongly implied Mars would be huge to the naked eye. What's funny is that the 2005 version of this email is exactly the same, except the dates were changed. The first email may have been an honest mistake by whoever wrote it, but the second was pretty clearly a practical joke. Yet a lot of people fell for it. It did the rounds again in 2006.

As for Mars, it continues on its merry way around the Sun. Sometimes, it's easy to think that planets can be smarter than people.

Bad Astronomy in Movies

Nobody cranks out bad astronomy like Hollywood. If the internet is a drinking fountain of science abuse, Hollywood is a gushing fire hydrant (no doubt knocked over during an exciting car chase which ends with a vehicle going over a cliff and exploding in flames).

The most highly-anticipated movie of the year was arguably 'Star Wars: Revenge of the Sith'. While there wasn't much astronomy in it, there was certainly a mish-mash of science. It had all the usual Hollywood stretch of truth: sounds in space (with no air to carry them), spaceships banking to make turns (with no air to push against), and gravity that is only around when you need it (a falling ship in space has people sliding around on it as the floor tilts).

But it also had one thing that bugs military strategists and science-fiction fans alike: using infantry to invade planets. Why do that, when you can stay safely in orbit and drop asteroids down? A 100-yard-wide rock falling at 20,000 miles per hour will do significantly more damage than a bunch of ill-programmed wise-cracking droids armed with buzzing flashlights. And once you're done softening up the ground troops with a few well-placed meteorite strikes, it's easy enough then to send down a mop-up operation.

Of course, this was a long time ago in a galaxy far, far away. Maybe the rules of engagement were different there.

But even on Earth in modern times, aliens use bad strategy. Just take a look at 'War of the Worlds': no, Steven Spielberg's remake of H.G. Wells' book, not the original 1953 movie version which kicks the new version's butt.

This movie doesn't suffer so much from Bad Astronomy as it does Bad But Convenient Script Writing.

Aliens (not martians: at least the writers figured out that wouldn't work too well) come to Earth riding, um, lightning bolts. They then control giant war machines a hundred yards tall that were hidden under ground, even though somehow a century of digging pipes, foundations, and subway systems somehow missed them (to be fair, one person explained this away by having the aliens grow the machines rapidly using nanotechnology—a real stretch, but a pretty nifty way to explain the movie's nonsense).

The super-lightning bolts fry all electronics so that cars don't work, though mysteriously a digital video camera is shown to be working just fine, evidently because it made the scene look cooler. Later, too, a news camera is seen to have caught the action as well. Maybe they have hand-cranked TV cameras in New Jersey.

Much fighting on the ground ensues (again, asteroids dropped from orbit seems like a better strategy). Eventually, the aliens succumb not to our weapons, which are useless, but to bacteria, against which they have no immunity. They die, Our Hero is reunited with his family, and all is well. Or Wells.

Speaking of which, go read Wells' original book. It's actually a page-turner, much better than any of the movie version of it, or even the rock-opera version (yes, there really is a rock-opera version). The book is extremely imaginative and far-reaching in its scope… though even Wells probably would have had a hard time seeing Tom Cruise playing the lead.

Bad Astronomy obituaries

Astronomy lost two popularizers last year, though at very different ends of the spectrum of reality. Bill Kaysing, the man credited with starting the conspiracy theory that NASA faked the Apollo landings, died in June 2005. He started writing about it during the Apollo missions, and stuck by his story until he died. The ridiculous idea that the Moon landings were hoaxed has been debunked countless times, but like a virus keeps popping up in different places. People will believe anything if it involves a government conspiracy. And while his theories were certainly wrong, he did force people to look at the Apollo missions again, and that maybe isn't such a bad thing: it was a magnificent achievement, and well worth a second look.

The other popularizer who died was Johnny Carson, in January 2005. Johnny was an amateur astronomer, possessed a telescope himself, and had many guests on his show who promoted science. He is clearly the man who launched Carl Sagan's career as a popular astronomer, and perhaps changed the way the public views scientists. Carson also had James 'The Amazing' Randi as a guest many times, who promoted critical thinking through magic tricks and debunking 'psychics' like Sylvia Brown and Uri Geller. Carson was much more than an entertainer, he was a science educator, who taught people how to think without them even knowing it., and they had fun the whole time. What better legacy can there be?

Bad Astronomy ... at NASA?

Bad science is pernicious, and it's everywhere. You might expect it to be festering in government somewhere, but at the nation's space agency?

Yes, even there. The story, which broke late in 2005, dealt with the Public Affairs Office at NASA. The role of the PAO is to act as a middleman between the science of NASA and the public. They write the press releases, contact the news media when stories come out, and generally create the public face of NASA. The vast majority of the time they do a great job.

But not always. Jim Hansen, a senior NASA scientist who works on climatology, has long been outspoken about anthropogenic (that is, caused by man) global warming. He has given many interviews about the topic, and doesn't shy away from speaking his mind.

So when a young man named George Deutsch, who worked at NASA's PAO, started telling Hansen he couldn't talk at some venues, Hansen unsurprisingly resisted. While it may be the PAO's role to decide to whom scientists should and should not speak, this is usually based on the science itself. However, it was clear that Deutsch's objections were based on political ideology, and not on a reasonable examination of the facts. Hansen was told he could not talk on National Public Radio, because, in Deutsch's own words, it was "the most liberal" media outlet in the country. Obviously, that shouldn't have anything to do with news, but it did make Deutsch's political motivations quite clear. In fact, Deutsch was a political appointee, given his job at NASA by the Bush Administration, even though he had no scientific or public affairs background at all.

A New York Times article broke this story in early 2006. Hansen claimed he was being muzzled by Deutsch for political reasons, and many scientists corroborated Hansen's story.

Furthermore, it was revealed that Deutsch had been doing what he could to force his own ideology into other areas of NASA science as well. A NASA webpage was being written about the origin of the Universe, and Deutsch told the author to add the word 'theory' every time the phrase 'Big Bang' was used. Certainly, the Big Bang is a theory, a scientific one. To scientists, that means that it is pretty much rock-solid fact, supported overwhelmingly by evidence, and the best model we have to explain the origin of the Universe.

But to the public, the word 'theory' means a guess, or a conjecture. It is hard to imagine Deutsch didn't know that when he demanded that the word be used.

In fact, it is clear that his purpose was exactly to undermine the scientific nature of the Big Bang. In an email to the website author, he said that the Big Bang is "not proven fact; it is opinion, [and] it is not NASA's place, nor should it be to make a declaration such as this about the existence of the universe that discounts intelligent design by a creator."

Yes, you read that right: Deutsch was injecting Intelligent Design, a discredited 'alternative' and decidedly religious idea about the origins of life and the Universe, into NASA science. To hammer this home, Deutsch went on: "This is more than a science issue, it is a religious issue. And I would hate to think that young people would only be getting one-half of this debate from NASA. That would mean we had failed to properly educate the very people who rely on us for factual information the most."

Even ignoring the obvious First Amendment issues here, it's outrageous that someone in the NASA administration would try to muddy good scientific analysis with demonstrably false ideas, and then claim he was the one trying to educate the public.

There are other stories of Deutsch's meddling with science, but it turns out he didn't last too long anyway: a blogger by the name of Nick Anthis (**http://scientificactivist.blogspot.com/2006/02/breaking-news-george-deutsch-did-not.html**) discovered that Deutsch lied on his resume, claiming he graduated from Texas A&M when in fact he dropped out to take the NASA job. A few days after that story broke, Deutsch resigned from NASA.

But the damage lingers on. A lot of scientists are pretty miffed about what happened. On top of that, NASA's 2007 budget contains devastating cuts to science, with many missions delayed indefinitely, or having their funding cut back, or being canceled entirely. This is a continuing story, and will no doubt have many more revelations as time goes on.

(Editor's note: See **http://www.nasa.gov** for the latest information regarding current missions and budget updates.)

NASA's job is threefold: explore the Universe with robots and people; perform top-notch, cutting-edge science; and to inform and excite the public about it. For decades it's done an outstanding job, but now all three areas are under fire, sometimes from within NASA itself. Hopefully the administration at NASA will work to correct this straying of its course.

Good Astronomy

There were more cases of bad astronomy, of course, there always are. Be vigilant! When you read an article, watch a movie, or hear a rumor about astronomy, check the sources, review the math, do a bit of web searching. Chances are you'll be able to find out the truth behind the fiction. In the end, Good Astronomy will win out.

Astronomer *Dr. Phil Plait* runs the web site **http://www.badastronomy.com**

The Cydonia region of Mars. The "Face on Mars" saga has become a classic of bad astronomy. It's another story that keeps popping up like a virus. While humans are pre-disposed to seeing the human form in anything from mountains to condensation on a building's window, the clearly natural formations on Mars suprisingly, even today, generate a following of people who cannot let go of the appeal of an artifically formed mountain built specifically for the benefit of inhabitants on the third planet from the Sun. The original Viking image is shown at left, and to the right is a 2006 image from the European Mars Express orbiter, revealing the geological detail of this fine mountain. Images courtesy (left) NASA/JPL and (right) ESA/DLR/FU Berlin (G. Neukum), MOC (Malin Space Science Systems).

11 Understanding GALAXIES and LARGE-SCALE STRUCTURES...

One of the great surveys of the cosmos using the Hubble Space Telescope is the COSMOS galaxy survey. *Anton Koekemoer*, an astronomer at the Space Telescope Science Institute, leads the observations, and here provides a complete background to the survey.

...Surveying the COSMOS with the HUBBLE Space Telescope

ONE OF the most profound questions in astronomy is how the galaxies and large-scale structures that we see in the universe today were assembled over the course of cosmic time. Coupled with this is the question of how the first stars and galaxies formed, and how they are related to the supermassive black holes that are inferred to be present at the centers of many galaxies that we observe today.

Galaxies can be generally classified according to their shape or morphology (spirals, ellipticals, irregulars, or interacting), their size, and their intrinsic brightness, or luminosity. In the local universe, galaxies are generally arranged in groups or clusters, with the richest clusters containing up to thousands of galaxies in regions up to several million light years across. The groups and clusters themselves are arranged in much larger filaments and sheets, all of which are interconnected with one another on scales up to several hundred million light years in extent.

There are a number of known relationships between the properties of galaxies and the types of environments in which they are found. The densest environments, such as those at the centers of very rich clusters, tend to have an excess of elliptical galaxies, while spiral and irregular galaxies are predominantly located toward the outskirts of clusters and in the low-density voids that lie between the cosmic filaments and sheets.

Furthermore, the largest, most massive, galaxies tend to be found in the densest, richest clusters, suggesting that they have been growing rapidly by accreting neighboring galaxies, while galaxies in less dense regions tend to be less massive.

Finally, there are known relationships between the properties of galaxies and their central supermassive black holes which are inferred to be present at their cores: when the masses of these black holes are estimated (ranging between 10 million

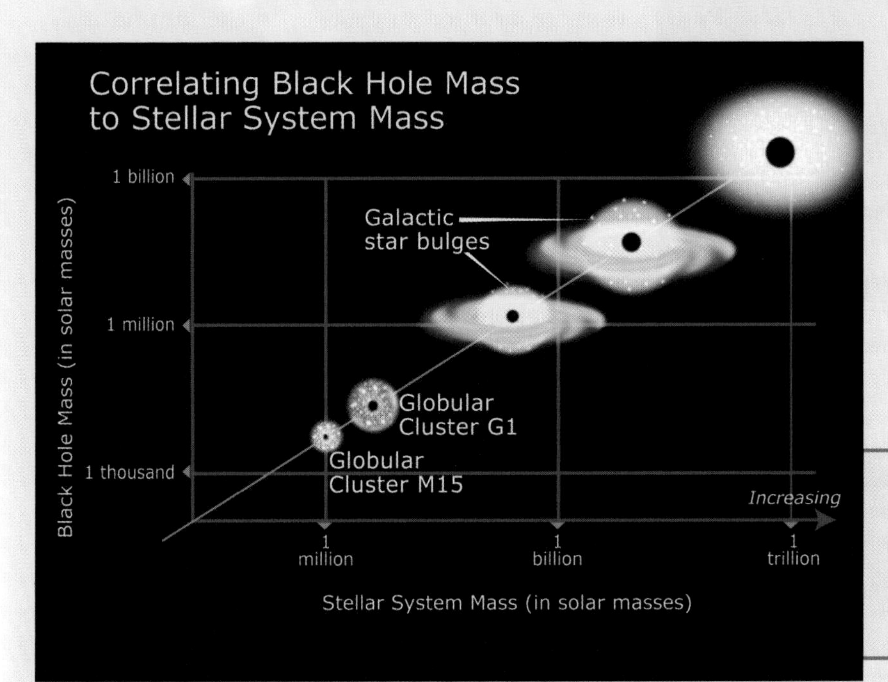

Right: *Measurements by the Hubble Space Telescope discovered the relationship between central black hole mass and the size of the star system. Image courtesy NASA and STScI.*

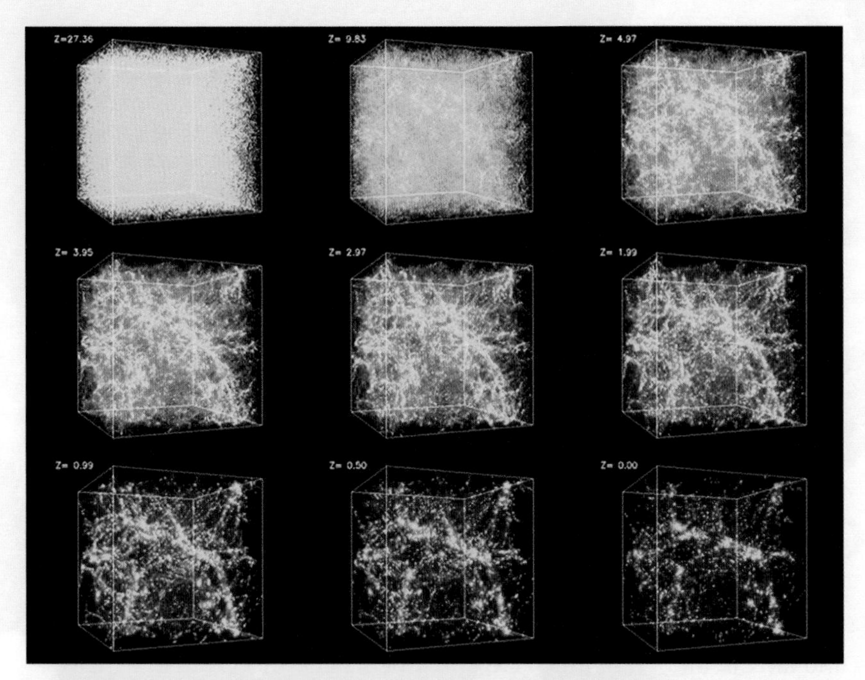

and one billion times the mass of our sun), it is found that the more massive black holes tend to be found in galaxies that are also more massive, suggesting a clear causal link between the formation and growth of galaxies and their central black holes.

Current theories of galaxy and structure formation suggest that at very early epochs (within the first 10 - 100 million years of cosmic time), slight excess density fluctuations in the primordial universe served as `seeds' which began to grow by accreting nearby matter through gravitational attraction. Detailed numerical simulations reveal an increasingly complex network of dense clusters and connecting filaments, growing throughout the course of cosmic time by attracting more surrounding material, while galaxies within these clusters and filaments also continue to grow and change with time.

The overall scheme of `hierarchical' galaxy formation describes the growth of galaxies as a continual process of accreting and merging with other galaxies. These theories and simulations make very specific predictions about the relative numbers of galaxies with different luminosities, i.e. the luminosity function, for the local universe today as well as for earlier epochs in cosmic time. In addition to changes in the luminosity function, these calculations also describe changes in the global rate of star formation and mergers between galaxies, thereby reflecting the interplay between the expansion of the universe and the gravitational collapse of material onto the filaments and clusters.

Although theories of hierarchical formation are able to quite successfully predict many of the relationships between galaxy properties and their environment that we observe in the local universe today, it is also necessary to test their predictions for the evolution of these properties with cosmic time, by observing samples of galaxies at much earlier epochs. This has historically been difficult to do with ground-based telescopes, since galaxies beyond about redshifts of 0.5 - 1 are generally not much larger than about 1 arcsecond, equivalent to the resolution limit usually achieved from the ground

Top: *Simulation of the formation of large scale structure in the universe, showing a 100 million light year cube at different epochs. At z=28, the universe is essentially homogenous. At later epochs, for example at a redshift of z=3, the filamentary structure of the universe is evident. At the nods of the filaments, superclusters of galaxies are forming. Image courtesy Andrey Kratsov, KICP.*

Above: *Three selected portions of the Hubble Deep Field taken in December 1995, and unveiled at the January 1996 meeting of the American Astronomical Society in San Antonio, Texas. Some of the dimmest galaxies ever seen, down to 30th magnitude, are revealed in these images. Image courtesy NASA and STScI.*

at optical wavelengths. However, at a redshift of 1 the universe is already relatively old at about 6 billion years; much of the more dramatic changes in galaxy evolution occurred at earlier epochs, and substantially better resolution is required to study these.

Probing the Distant Universe

The advent of the Hubble Space Telescope provided a significant improvement of at least a factor of 10 in spatial resolution, thus opening up the early universe by providing uniquely detailed images of distant galaxies. The original Hubble Deep Field North, using the Wide Field / Planetary Camera 2 (WFPC2) on Hubble, provided the first deep look at the distant universe. It achieved a depth of about 28th magnitude in four filters (centered at 300nm, 450nm, 606nm and 814nm) with a pixel scale of 0.04 arcseconds, detecting about 4,000 galaxies, many of which are at redshifts 2 - 3 and extend up to redshifts of about 5. This provided the first view of the time evolution of quantities such as the cosmic star formation rate. This peaked somewhere between redshifts 1 - 2 (when the age of the universe was between about 3 and 6 billion years). The rate has subsequently declined significantly to the present time.

While the depth and spatial resolution of the Hubble Deep Field North enabled the first direct detections of galaxies beyond redshifts 2 - 3, the actual area covered by the field was relatively small, extending only about 2 arcminutes on a side (or 1/15th the diameter of the full Moon). At redshifts of 2 - 3, this corresponds to a length scale of about 3 million light years across the entire field, which is not much larger than the size of a single cluster in the local universe today, and certainly much smaller than the size scale of cosmic filaments and sheets which can extend for hundreds of millions of light years.

This introduces the problem of `cosmic variance', in which a very small field such as the Hubble Deep Field North might very likely not probe a representative volume of the universe, but may instead by chance land on a region that is either over-dense or under-dense, and thereby not give a complete picture of galaxy evolution over cosmic time.

One way to mitigate cosmic variance is to obtain a completely different sightline through the universe, and for this purpose the Hubble Deep Field South was carried out. This indeed revealed that there were very significant differences in the cosmic structures probed by these two narrow lines of sight, which would bias the results obtained from any single field of such a small size.

The GOODS Survey

The most effective means of mitigating cosmic variance is to cover a much larger area. This was made possible with the installation of the Advanced Camera for Surveys (ACS) on board Hubble, which is several times more sensitive than WFPC2, particularly at the red end of the spectrum, while at the same time covering a larger area with improved spatial resolution.

This camera was used in the Great Observatories Origins Deep Survey (GOODS) to cover two areas of sky in the North and South with a total area about 33 times larger than the combined area of the original Hubble Deep Fields North and South, while achieving a similar depth of about 28th magnitude. Four filters were also used for this survey, centered at 475nm, 606nm, 775nm and 850nm, thus utilizing somewhat redder

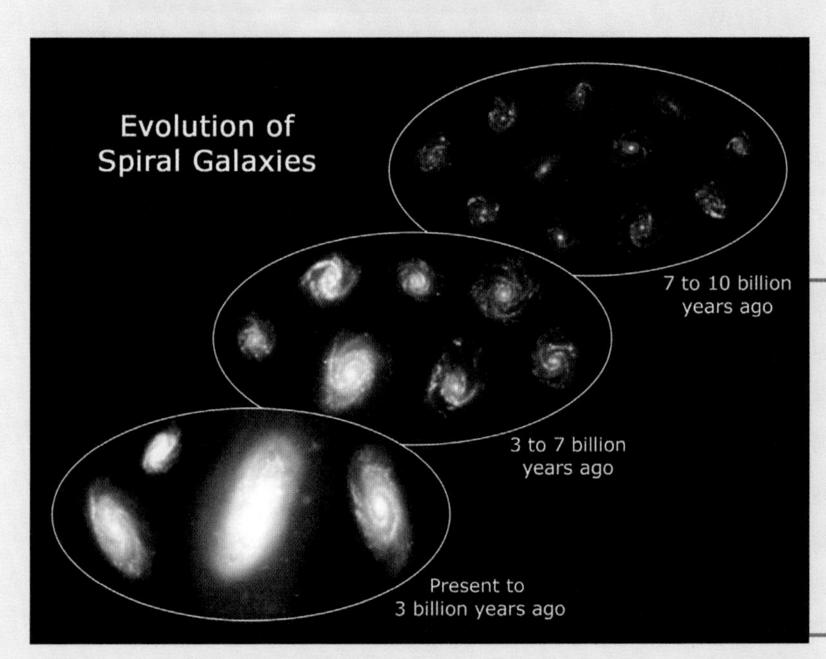

Left: These galaxies in this montage were captured in the Great Observatories Origins Deep Survey (GOODS). Fully developed galaxies are found up to 3 billion light years away. Between 3 to 7 billion years galaxies the spiral structure is not as well developed. Weakly defined spiral arms feature in the galaxies located between 7 and 10 billion light years away. Because it takes light a finite tie to travel to us from across the universe, we see these galaxies as they were in the past. Image courtesy NASA and STScI.

wavelengths than the Hubble Deep Field to better facilitate exploration of the high-redshift universe.

The most important feature of the GOODS survey is its combination of red sensitivity and area, which not only covers many more lines of sight than the Hubble Deep Fields, but in addition provides a sample of about 40,000 galaxies.

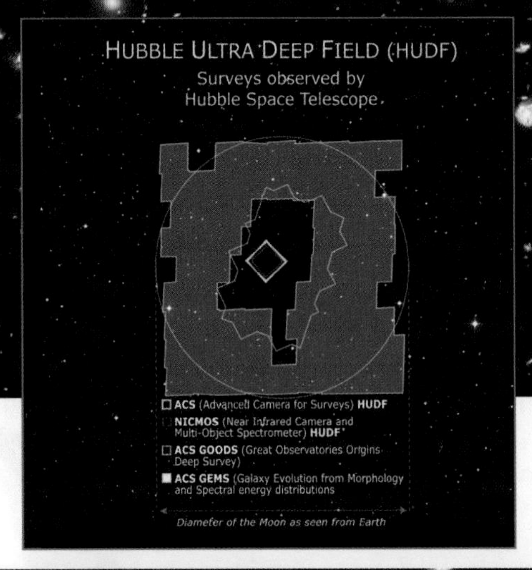

Top: *The Hubble Ultra Deep Field is the deepest ever optical light image of the universe, taken by the Hubble Space Telescope between September 2003 and January 2004. The image stacked 800 images taken over 400 orbits of Earth. Nearly 10,000 galaxies, most never seen before, are revealed in the HUDF. Image courtesy NASA, ESA, S. Beckwith (STScI) and the HUDF Team.*

Middle: *A comparison of the field of view for the Hubble Ultra-Deep Field, GEMS and GOODS surveys. The circle represents the diameter of the Full Moon. Image courtesy NASA, A Field, and Z. Levay (STScI).*

Bottom: *A small region of the HUDF reveals a distant spiral galaxy and dozens of more distant galaxies. The smallest objects in this image come from light from galaxies that formed just one billion years after the big bang. Image courtesy NASA, ESA and B. Mobasher (STScI/ESA).*

More than a thousand are above redshift 3, and several hundred galaxies extend up to redshifts 5 - 6 when the universe was only about 1 billion years old.

In addition, these fields have extensive multi-wavelength coverage ranging from X-rays through to radio wavelengths, which reveal almost 1,000 active galactic nuclei, some of which extend up to at least redshift 6, and possibly further. The active galactic nuclei are inferred to be supermassive black holes that are being fuelled for some period of time by accreting gas, very often the result of mergers or interactions between galaxies. Consequently they provide direct tracers of the frequency with which galaxies interact or merge.

The large numbers of high-redshift galaxies from GOODS, combined with the extensive area covered, has enabled a determination of the rate of cosmic star formation beyond redshift 2, all the way up to redshift 6, showing that the rate of star formation appears to have remained approximately constant during this period. This means that, at least since the time when the universe was about 1 billion years old, galaxies have been growing by star formation at a more or less constant rate, until the universe reached an age of around 6 billion years, after which the rate of new star formation declined significantly. In addition, however, the relatively flat rate of star formation up to redshift 6, combined with the fact that the universe appears to be transparent, or `reionized', up to at least redshift 6, implies that stars and galaxies must have begun forming at much earlier epochs.

Going Deeper than 'Deep'

In order to probe the populations of galaxies around redshift 6, deeper imaging was needed since the GOODS data are only sensitive enough to probe the brightest end of the galaxy luminosity function at these redshifts. As a result, in 2004, the Hubble Ultra Deep Field (HUDF) was obtained. Hubble observed a single piece of sky using the same four filters as GOODS with a total exposure time of 1 million seconds.

The HUDF reached about six times fainter than the HDF, down to almost to 30th magnitude. Although this is a smaller field of view than GOODS, it was placed within the GOODS area to determine the effects of cosmic variance. The increased sensitivity allows for the study of much fainter galaxies at

■ *The COSMOS image relative to the Full Moon. Image courtesy Jean-Paul Kneib (Laboratoire d'Astrophysique de Marseille).*

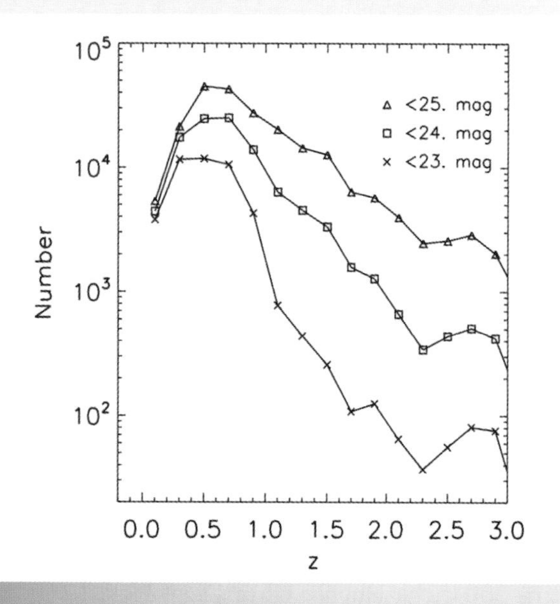

Above: *A graph showing the number of galaxies at different redshifts (z) using results from the COSMOS survey. The bottom curve shows galaxies fainter than 23rd magnitude; the middle curve shows galaxies fainter than 24th magnitude; the upper curve shows galaxies fainter than 25th magnitude. Image courtesy N. Scoville and the COSMOS team.*

Below: *The distribution of dark matter on large scales is revealed in this single frame from the Millennium Simulation, the largest N-body simulation carried out thus far (more than 10^{10} particles). The filaments and nodes are locations where clusters of galaxies form. The linear scale represents the same image scale as the COSMOS survey. Image courtesy Springer (2005).*

redshift 6, thereby sampling enough of a range in luminosities to enable the construction of galaxy luminosity functions at redshift 6. This revealed that the luminosity function at redshift 6 is remarkably similar to that at later epochs, and more specifically, that the number of faint galaxies at these redshifts are not enough to have reionized the universe. The reionization had to have happened at earlier epochs.

Probing to earlier epochs when reionization may have occurred requires observations of the universe at longer wavelengths. This is because the light from galaxies becomes shifted to gradually longer wavelengths at increasingly high redshifts; by redshift around 6, for example, the reddest optical filters (around 850nm) probe light that is actually emitted with a rest wavelength of around 120nm, far into the ultra-violet. Because of the large amounts of intervening absorbing material, very little light at shorter wavelengths actually reaches our telescopes. This results in objects at redshift 6 or above not being generally detectable at any wavelengths shorter than about 850nm. However, longer wavelength emission from sources at such redshifts is readily detectable by using infrared detectors.

For example, observations with the Near Infrared Camera / Multi Object Spectrometer (NICMOS) on board Hubble can sample wavelengths up to 2.2 micron, while the Spitzer Space Telescope is able to observe at wavelengths between 3.6 and 8 microns in one camera, and 24 to 160 microns in another of its cameras. Similarly, the James Webb Space Telescope, expected for launch after around 2011, will cover similar wavelengths but with about a factor of 100 increase in both resolution and sensitivity.

At redshift 6, an observed wavelength of 3.6 microns corresponds to a rest wavelength of about 500nm. With JWST sensitivity ranging from 3.6 to 8 micron, it can observe galaxies at higher redshifts, reaching up to redshift 12 or beyond, when the universe was only about 370 million years old. These timescales become comparable to the time taken for a single galaxy to form, which means we are really starting to be able to probe the era of formation of the very first stars, galaxies and black holes.

The COSMOS Survey

A complementary approach to surveys such as the Hubble Ultra Deep Field is to carry out a survey with the widest possible area using the Hubble Space Telescope. The COSMOS project has achieved this, spending a total of 600 orbits (1.5 million seconds, the largest amount of time ever awarded on Hubble to a single project) to map out a section of sky covering a total of 2 square degrees, or about 1.5 degrees on a side.

The linear extent of the survey is about 3 times larger than the diameter of the full moon, and the area could contain 9 full moons stacked in a grid next to each other. The aim of this project is to sample the largest possible cosmic structures at a redshift of about 2, when the structures were likely becoming fully developed. At this redshift, the field covers over four hundred million light years, much larger than the largest known cluster, and comparable to the lengths of the filaments and sheets that connect the clusters.

Three Million Galaxies

A total of about 3 million galaxies have been identified in this survey to date, most of which are at redshifts 1 - 2, although several thousand may extend up to redshifts of 5 or 6. The field also has excellent coverage in the X-rays with the XMM/Newton telescope, and at infrared wavelengths with the Spitzer Space Telescope, thereby allowing the identification of thousands of active galaxies.

Preliminary results from this project indicate that large-scale structure at redshift 2 indeed appears to be qualitatively different from that in the local universe, suggesting that the largest structures in the universe continue to evolve and change on much later timescales than the galaxy population, which is already well in place at earlier epochs.

A massive campaign using numerous telescopes around the world is following up the Hubble survey. Many telescopes are being used to measure every possible attribute to each object in the COSMOS field, including redshift. The XMM Newton observatory is following up with detailed X-ray maps, and Galex is doing the same with ultraviolet imaging. The Spitzer Space Telescope is targeting infrared sources, and additional observations by the Chandra X-ray observatory complete the space-borne activities. Some of the largest telescopes on Earth are also joining the campaign.

The 8-m Subaru telescope and the Canada-France-Hawaii telescope are performing multicolor imaging, with additional involvement from the UK Infrared Telescope and the 10.4-m Caltech Submillimeter Observatory on Hawaii. The Very Large Array in Socorro, New Mexico performs radio wavelength imaging, and Europe's Very Large Telescope and the 8-m Magellan telescope performs spectral observations of more than 50,000 galaxies to determine redshifts.

Such a detailed campaign concentrating on one large area of the sky will help characterize the shape and structure of galaxies at different distances, and hence different epochs, on a scale that encompasses the large scale structure of the universe. Only tiny portions of these structures were viewed in all previous deep sky surveys.

With the advent of new telescopes in the next decade, such as the Atacama Submillimeter Array, the Giant Magellan Telescope, and the James Webb Space Telescope, COSMOS survey and others that will follow are beginning to piece together an accurate picture of the real universe. One important aspect of this research is to match what is predicted by supercomputer simulations using models of the Big Bang that build the large scale structure using various values for dark matter, expansion rates, dark energy, with the real observations.

12 The most POWERFUL EXPLOSIONS in the Universe...

Neil Gehrels, Principal Investigator for the SWIFT gamma-ray observatory, with co-author and colleague, *Peter Leonard,* outline some dramatic new discoveries and the progress made in the past year in our understanding of the origin of some of these bursts

...Gamma-Ray Burst DISCOVERIES with the SWIFT MISSION

THE DRAMA of 18 February 2006, will be remembered for a very long time by astronomers working with the Swift gamma-ray observatory, adding to the growing list of exciting times in gamma-ray astronomy. On that memorable day, a cosmic gamma-ray burst that later turned out to be the second closest recorded event to date was detected. That alone would be reason for excitement. But this burst was different in a number of ways. Its uniqueness represents exactly the kind of object the Swift spacecraft was built to detect, and the quick follow-up observations that were triggered by Swift will provide valuable data helping to unravel one of the longest standing mysteries in astronomy today.

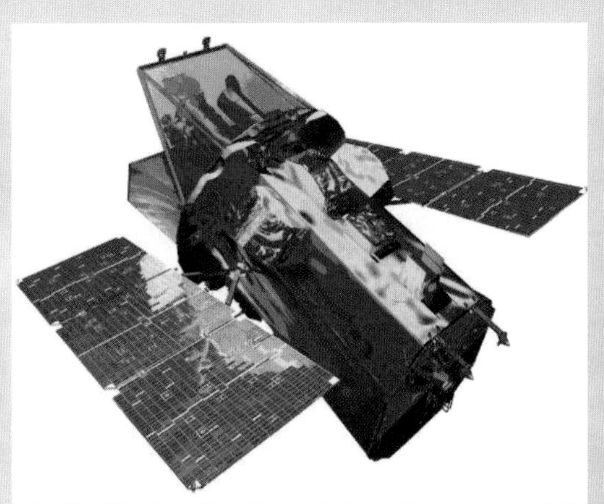

Swift is a multi-wavelength space-borne observatory that detects cosmic gamma-ray bursts, quickly determines the precise location of the bursts in the sky (i.e. localizes), and promptly provides high-resolution follow-up observations. Swift was launched into space on 20 November 2004, and is carrying out measurements of gamma-ray bursts and their afterglows with unprecedented detail. A key result from the first year of Swift operations is that "short" gamma-ray bursts have a different physical origin than the "long" bursts, and are likely caused by mergers of pairs of compact objects (i.e., neutron stars and/or black holes). Swift has also provided observations of gamma-ray bursts at high redshift, a giant flare from a soft gamma repeater, continuous precision monitoring of a Type Ia supernova, and the beginnings of an unbiased catalog of active galactic nuclei.

Gamma-Ray Bursts

Cosmic gamma-ray bursts (GRBs) are the most powerful explosions in the Universe since the Big Bang. They occur at random times and locations in the sky, and last anywhere from a fraction of a second to several minutes. A GRB in progress is typically the brightest gamma-ray source in the entire sky. After the burst is finished, it is never seen again at gamma-ray energies.

The big breakthrough in GRB research was the discovery of X-ray "afterglows". While the gamma-ray component of a GRB is typically very fleeting, there is often lingering X-ray emission from the burst region. X-ray telescopic observations of this afterglow can pinpoint the GRB location on the sky. This information is then disseminated as widely and quickly as possible, to allow high-resolution, ulti-wavelength, follow-up observations of the burst locations with a variety of ground - and space-based observatories. This now routinely leads to the identifications of GRB host galaxies.

Above: *The Swift gamma-ray burst observatory. Image courtesy Swift Team.*

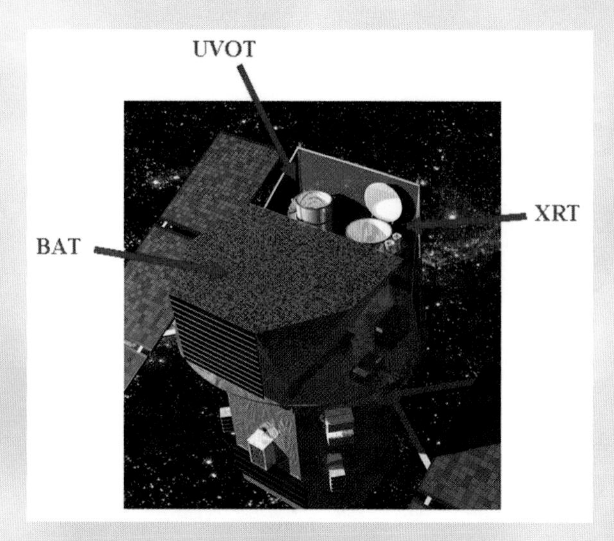

This technique was pioneered in 1997 thanks to the Italian-Dutch BeppoSAX satellite, and the conclusion was reached that the long GRBs (a class of bursts that typically last more than two seconds) result from enormous explosions of massive stars in distant galaxies. Specifically, the widely accepted picture is that a core collapse of a rotating massive progenitor star produces a black hole, which results in a pair of oppositely directed "jets" of matter being ejected along the rotation axis of the material that is being accreted onto the black hole. If either of these narrow relativistic particle jets is pointed toward the observer, then a GRB is observable. A supernova explosion is sometimes detectable many days later.

While BeppoSAX certainly had a built-in design capability to observe GRBs, the satellite was not optimized for this purpose. In particular, there was a delay of hours between when BeppoSAX's wide-field X-ray camera first detected a burst, and when the satellite's high-resolution X-ray telescopes could be pointed at the burst location to carry out follow-up observations and provide a refined the position for the burst. Consequently, the dissemination of an accurate GRB position in the sky to other astronomers at major observatories did not happen in a manner that was optimum for such a rapidly fading phenomenon.

This was a problem especially for the short GRBs, which is a class of bursts that typically last less than two seconds. The short bursts contain a higher proportion of high-energy gamma-rays compared with the long bursts. These temporal and spectral differences led many scientists to suspect that the short bursts likely have a different origin compared with the long bursts.

NASA's small explorer satellite named HETE (High Energy Transient Explorer) enjoys a significant improvement over BeppoSAX in turnaround time. HETE was launched on 9 October 2000, and was the first satellite designed primarily to observe GRBs. Once the ground system was perfected, HETE began to disseminate accurate GRB locations world-wide via the Internet in seconds. One limitation of the HETE satellite is that it carries only small instruments, and it cannot be re-oriented to better observe a burst.

Swift is the next step up from HETE, namely a mid-sized Explorer satellite with larger instruments and an autonomous re-pointing capability. The primary purpose of Swift is to "swiftly" localize cosmic gamma-ray bursts (GRBs) in the sky, and to re-orient itself in several tens of seconds so that the spacecraft's high angular resolution instruments can observe the burst in great detail for an extended period of time. Swift also quickly disseminates all the key information regarding the burst world-wide in seconds via the Internet to allow rapid follow-up observations of the burst location via ground- and space-based instruments.

Spacecraft Capabilities

Swift carries three scientific instruments: the Burst Alert Telescope (BAT), the X-Ray Telescope (XRT) and the UltraViolet/Optical Telescope (UVOT). The BAT is a wide-field instrument that makes the initial detection of a GRB. This

Above: *The Swift gamma-ray burst observatory includes three telescopes: Burst Alert Telescope (BAT), X-Ray Telescope (XRT), and UltraViolet Optical Telescope (UVOT). Image courtesy Swift Team.*

Right: *Dr. Neil Gehrels displays a section of the coded aperture mask for BAT and one of the detectors for Swift in his office at the Goddard Spaceflight Center. Image courtesy Martin Ratcliffe.*

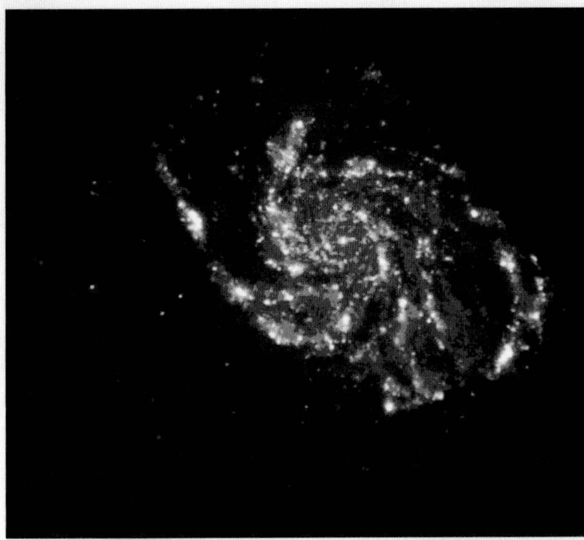

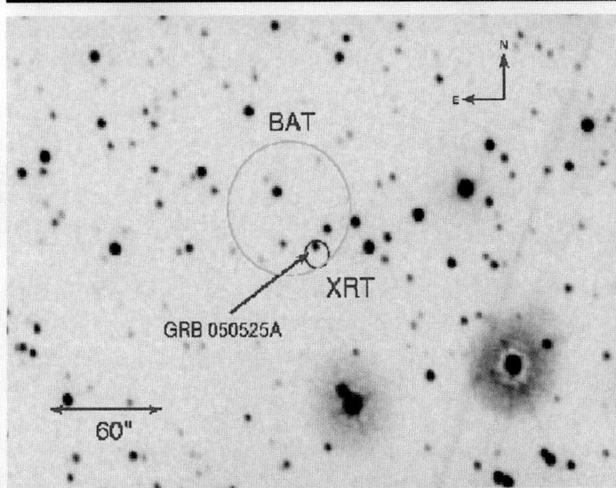

information is then used to rapidly realign the spacecraft to allow the XRT and UVOT to scrutinize the burst location in great detail. The latter two instruments are co-aligned and have relatively narrow (roughly 20 arc minute) fields of view.

The BAT is the largest instrument aboard Swift. It boasts 32,768 CdZnTe (Cadmium-Zinc-Telluride) detectors, each 4 mm by 4 mm by 2mm, which sit in the "focal" plane behind a coded-aperture mask. A coded-aperture mask is like a pinhole camera, except instead of having only one hole, there are many holes arranged in a mathematical design. The pattern of light that reaches the focal plane through the mask is used to accurately reconstruct an image of the source (or sources) in the sky. The mask consists of more than 50,000 lead tiles, each 5 mm by 5 mm by 1 mm.

The BAT continuously observes more than one-third of the observable sky (i.e., the sky not blocked by Earth) for GRBs and other high-energy transients. The BAT quickly detects any such transient, and provides a position to better than two arc-minutes. The spacecraft then reorients itself to put the burst within the XRT and UVOT fields of view.

The sensitivity of BAT to GRBs is about 2.5 times higher than that of the highly successful BATSE (Burst And Transient Source Experiment) instrument on CGRO (Compton Gamma-Ray Observatory). BATSE was an all-sky monitor that detected more than 2600 GRBs from 1991 to 2000. The BAT on Swift is detecting roughly 100 GRBs per year, which is approximately the rate expected.

The XRT consists of a 'Wolter 1' grazing-incidence telescope that focuses X-rays onto a single CCD. The latter is a state-of-the-art "CCD-22", the same as those designed for the NASA-ESA XMM-Newton mission that was launched in 1999. The CCD is sensitive to X-rays with energies in the range from 0.2 to 10 keV with excellent quantum efficiency at the low-energy end.

The XRT detected more than 90% of the GRBs pointed at during the first year of Swift operations. The XRT is a several orders of magnitude more sensitive than the BAT, and has proven to be about five times more sensitive than the best instruments aboard the now defunct BeppoSAX spacecraft. The XRT can follow the decline of a GRB afterglow for many days.

The field of view of the XRT is 23.6 arc-seconds by 23.6 arc-seconds, and positions of GRB afterglows accurate to about 6 arc-seconds are provided quickly. Positions accurate to about 2 arc-seconds are derived after careful reprocessing.

The UVOT is a modified Ritchey-Chretien design with an aperture of 30 cm, and is based upon the Optical Monitor aboard XMM-Newton. A microchannel-plate-intensified CCD detector that is sensitive to photons in the range from 170 to 650 nm sits in the focal plane. In front of the detector there is an 11-position filter wheel containing grisms and various filters to allow both low-resolution spectrometry and broadband photometry. The UVOT has two of these detector systems for redundancy. Either one can be selected via a steerable mirror mechanism.

The UVOT detected GRB counterparts in about 30% of the pointing opportunities during

Above: *Two examples of UVOT images: spiral galaxy M101 (upper), and GRB 050525A (lower). Image courtesy Swift/UVOT Team.*

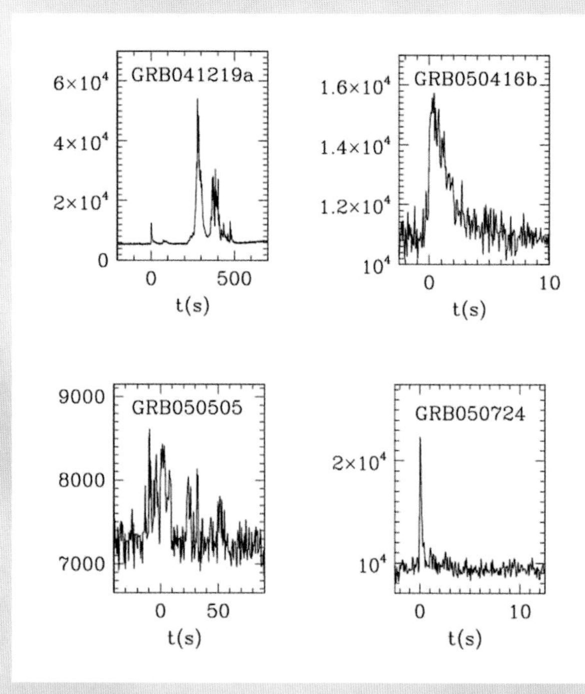

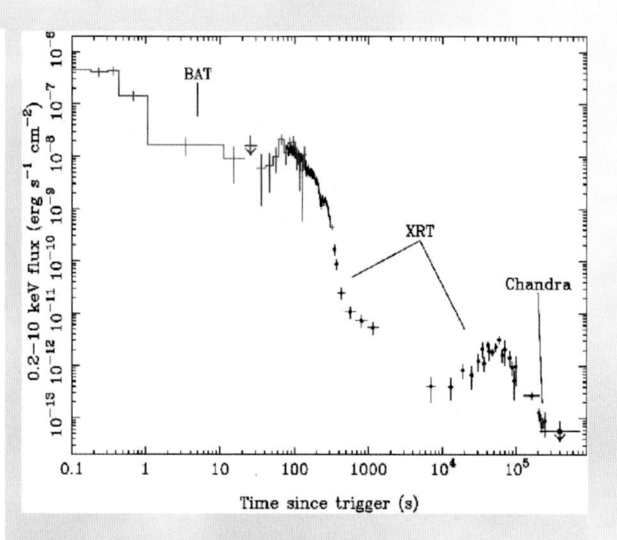

the first year of Swift operations. The UVOT can provide an image down to a visual magnitude of 19 in less than 1 hour, and to magnitude 22 in about 3 hours. Positions accurate to 0.33 arc-seconds are derived from UVOT observations.

The Discoveries

Swift has produced a string of discoveries during its first year in orbit that have fascinated astronomers. Several short bursts have been localized and we now believe these are the birth events of black holes in the fiery mergers of pairs of compact stars, a different physical mechanism than for the long bursts. A GRB was detected from an extremely large distance, becoming one of the most distant objects observed by any means. Observations of the afterglows of GRBs have found a bright early component and huge flares, both pointing to a new understanding that the prompt phase of bursts lasts for hundreds of seconds instead of the tens of seconds as thought before. In non-GRB science, a gamma-ray flash was discovered from a galactic magnetic star that was brighter than any transient emission ever seen. The first survey of the sky at wavelengths between X-rays and gamma-rays has found that black holes at the center of active galaxies are often surrounded by obscuring clouds of gas. Swift has also provided photometric monitoring of a Type Ia supernova with unprecedented accuracy, especially in the ultraviolet.

■ ***Above left**: Four examples of BAT observations of the prompt emissions of gamma-ray bursts. The vertical axis is the count rate in counts per second. The horizontal axis is time since the burst trigger in seconds. Image courtesy Swift Team.*

■ ***Above right**: The combined BAT/XRT/Chandra light curve of the X-ray afterglow of GRB 050724. The vertical axis is the energy flux in ergs per square centimetre per second in the 0.2 to 10 keV energy range. The horizontal axis is time since the burst trigger in seconds. The BAT data points are extrapolations in the X-ray band based on the measured spectrum. Image courtesy Scott Barthelmy et al.*

Short GRBs

Prior to Swift, no short GRB had been accurately localized, and no afterglow emission from a short GRB had been detected.

Swift rapidly localized seven short GRBs to arc-second accuracy during the first year of operations, and two of these bursts have been convincingly associated with host galaxies. GRB names are determined by the date on which the burst occurs, so an event on 2005 May 9 becomes GRB 050509. If two bursts occur on the same date, then a letter follows the name.

GRB 050509B was the first short burst to be localized accurately by Swift, and it was also the first short burst ever for which an X-ray afterglow was detected. This burst was very weak (just detectable with the BAT) and very short (30 ms). The spacecraft slew was completed in 52 seconds, and the XRT revealed a very faint fading source. The location put it on the outskirts of an early-type (elliptical) galaxy at a redshift of 0.225.

GRB 050724 was a considerably stronger short GRB that was detected and localized by Swift. An excellent X-ray light curve was captured by over a period of 100 hours by the BAT, the XRT, and, ultimately, the Chandra X-ray Observatory, complete with two X-ray flares (described below). This GRB was associated with another early-type (elliptical) galaxy, this time at a redshift of 0.258. It also had an optical and radio afterglow.

GRB 050709 was a short GRB that was initially discovered and localized by HETE, and was later observed by Swift and many other observatories. It was the first short burst with a

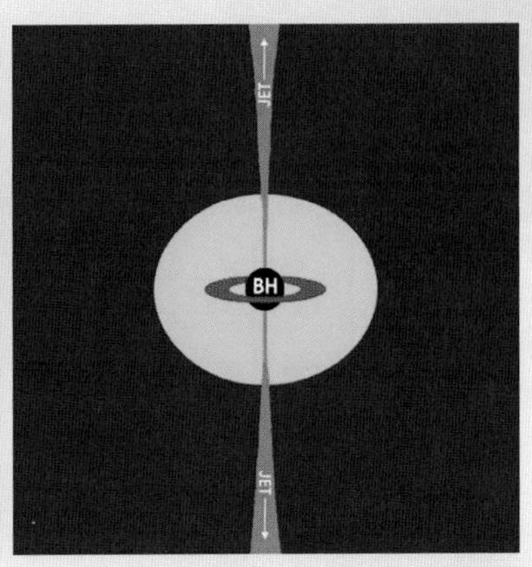

detected optical afterglow. Unlike the short GRBs localized by Swift, the July 9 burst was associated with a late-type (star forming) galaxy at a redshift of 0.160. However, this burst did not occur in an obvious region of star formation within the galaxy, which is distinctly different from what is invariably found to be the case when a long GRB is localized within a resolvable galaxy.

All three of these bursts support the hypothesis that the short GRBs originate from mergers of stellar-mass compact objects. Such objects include neutron stars and black holes. Neutron stars and black holes are the dense remnants of the cores of massive stars, and are produced during a supernova explosion. These compact remnants continue to exist long after all traces of their natal supernova events are gone. As billions of years pass, these objects drift far away from their regions of birth. These remnants are present in non-star-forming galaxies as relics of ancient episodes of star formation.

The majority of compact remnants are solitary objects, but it is an observational fact that binary pairs of compact objects exist in our Galaxy, and, in all likelihood, throughout the Universe. There are a variety of well defined scenarios for the creation of such pairs of stellar remnants, which is beyond the scope of this article.

Every compact-object pair in the Universe is destined to coalesce, primarily as the result of the emission of gravitational radiation. Gravitational waves are generated as the component objects orbit each other, and the energy source for these waves is a decrease in the orbital energy of the pair. The pair moves closer together as each bit of orbital energy is transmitted into space in the form of distortions of space and time. The rate of energy emission depends very strongly on the separation of the pair, and increases rapidly as the separation decreases, which ultimately leads to a violent merger of the two objects. If both objects are neutron stars, then the merger creates a black hole. If either or both of the merging objects are black holes, then an even more massive black hole is formed.

The time scale for the merger of a pair of compact objects due to the emission of gravitational radiation depends on the component masses and their separation. Any realistic population of compact binaries will have various component masses, and a wide range of orbital separations. This will result in a broad spectrum of merger time scales. The mergers that occur in the present Universe have typically been in the works for hundreds of millions to billions of years. Hence, any given pair of compact objects will have likely drifted within its host galaxy for a very long time before the inevitable merger takes place, and, therefore, the resulting coalescence event is quite unlikely to be associated with any region of star formation within the galaxy, unless by random chance.

The favored source of the long GRBs, namely a hypernova/collapsar origin, is essentially ruled out for the short bursts. The collapsar-forming events are the supernova explosions of massive stars, and such supernovae do not occur in early-type galaxies, because such galaxies do not contain any massive stars. The long GRBs are observed to occur only in late-type galaxies, and these bursts are invariably associated with obvious regions of star formation within such galaxies. This is quite unlike the HETE short burst, GRB 050709. Therefore, a collapsar origin is essentially ruled out for the short bursts.

While the physical process that leads to the production of the black hole in the short-GRB process is quite different from that in the long-GRB process, the mechanism for the emission

■ ***Above**: Hypernova/Collapsar scenario. (left) Core of massive star collapses to a neutron star. (middle) Neutron star accretes from the disk and eventually collapses to a black hole (or a black hole forms rapidly in step a). (right) Particle flow from the poles of the black hole, and bore out of the surrounding gaseous envelope as the black hole accretes matter.*

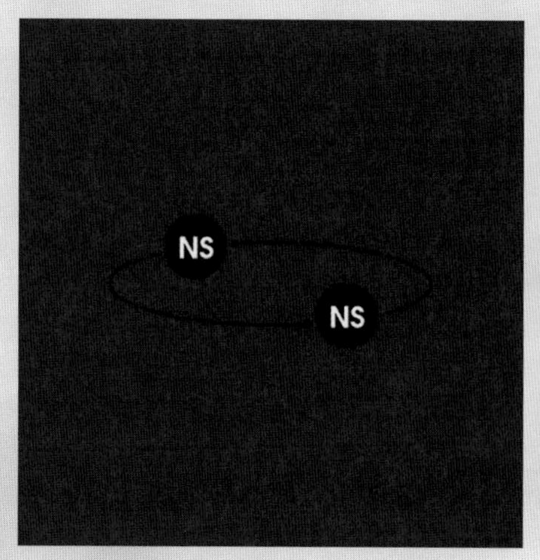

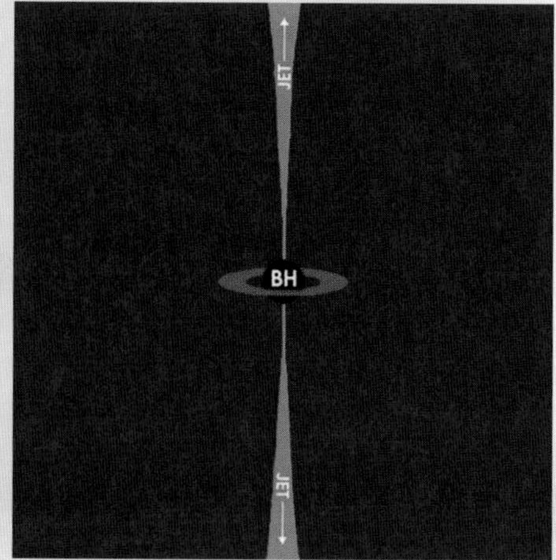

of gamma-ray photons is very similar in the two cases. Specifically, in both cases the prompt burst emissions result from the ejection of a pair of relativistic particle jets from near the poles of a black hole fueled by an accretion disk.

One difference between the two cases is that the accretion disk is likely to be much more massive for the long bursts, and this larger amount of fuel can explain why the long bursts last longer than the short bursts.

Another difference between the two processes is the environment. The long bursts occur in star formation regions full of gas and dust, which slow down the jet, and interfere with the escape of photons. In comparison, the short bursts occur in relatively clean environments, and this allows the higher energy gamma-rays to escape more easily. This can explain the harder spectra of the short bursts compared with the long bursts.

Swift also detected the short burst named GRB 050813, which had a duration of 0.6 seconds, and was very dim. This burst is possibly associated with a faint host galaxy at a surprisingly high redshift of 1.8.

Many more Swift observations of short GRBs are anticipated, and these should help clarify the situation further.

High Redshift GRBs

Swift is detecting more distant GRBs than both BeppoSAX and HETE due to the superior sensitivity of the BAT. Indeed, Swift appears to be measuring the intrinsic GRB redshift distribution, which is expected to have a peak at a redshift of two based on our current understanding of the star formation history of the Universe.

The current record holder for most distant bursts is GRB 050904, which was localized by Swift, and its coordinates were provided around the world within minutes. Ground-based follow-up observations discovered an optical afterglow, and measured a redshift of 6.29, smashing the old record for a GRB of 4.5. A redshift of 6.29 corresponds to a distance of 12.8 billion light years, which makes GRB 050904 one of the most distant objects in the Universe detected by any means.

The table below shows other distant Swift bursts with measured redshifts:

Name	Redshift
GRB 050505	4.27
GRB 050814	5.3
GRB 060223A	4.41
GRB 060510B	4.9
GRB 060522	5.11

In response to the Swift observations, it has been theorized that the low metallicity of the

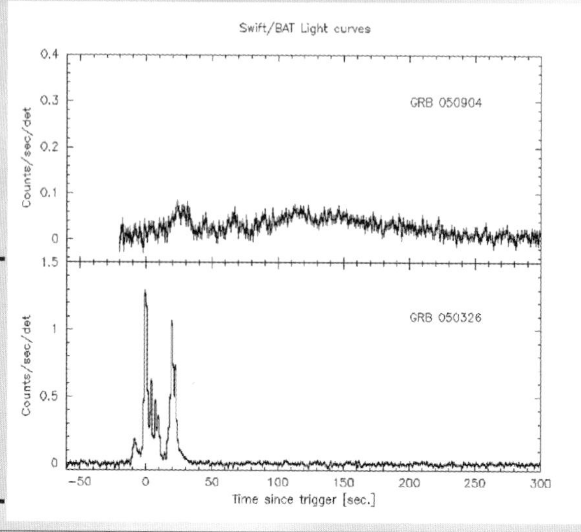

Above: *(Left) A double neutron star binary spirals together due to emission of gravitational radiation. (Right) Neutron stars merge and collapse to a black hole surrounded by a disk, with particle outflows emanating from near the poles.*

Right: *BAT light curves for the high-redshift gamma-ray burst GRB 050904 (upper), and a burst at a typical redshift (lower). The vertical axis is the count rate in counts per second per detector. The horizontal axis is time since the burst trigger in seconds. Image courtesy Swift Team.*

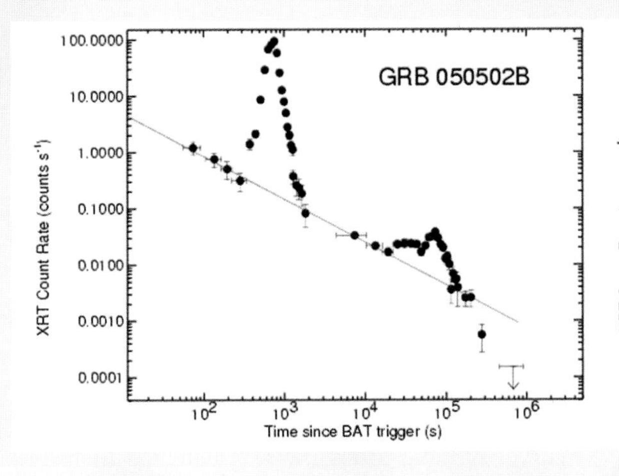

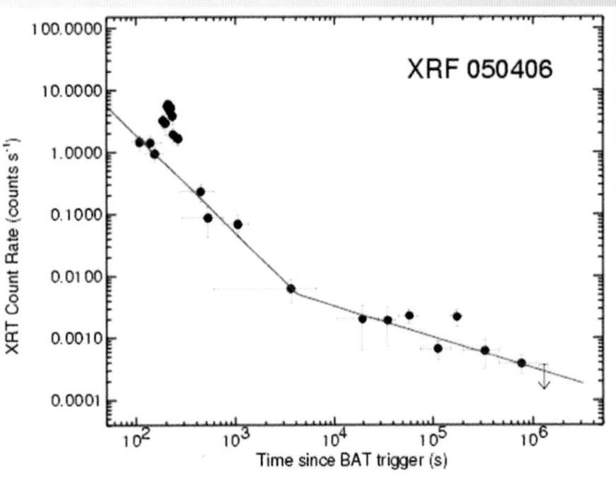

early Universe may have preferentially produced the high luminosity GRBs that are detected at high redshift.

Flares in GRB Afterglows

Swift has opened the door for detailed analyses of GRB X-ray afterglow light curves, which constrain many aspects of the physics of the so-called "fireball". This includes the discovery of two entirely new phenomena in burst afterglows: a bright early phase, and "flares". The latter have been described as "secondary explosions", "delayed outbursts", and even "hiccups".

Swift first detected flares in X-ray flash XRF 050406 and GRB 050502B. (An X-ray flash is a GRB-like event that is rich in X-rays.) The phenomenon was especially striking in GRB 050502B, where the prompt gamma-ray emission lasted 17 seconds, and 500 seconds later there was clearly a second X-ray event. An additional flare was observed after 10,000 s. Careful examination of Swift XRT observations of GRBs reveal that flares occur in roughly one third of observable X-ray afterglows.

The presence of the flares suggests that the GRB central engine which creates the fireball that produces the prompt gamma-ray emission is still active for thousands of seconds after the onset of the burst.

Flares seem to be common among bursts at high redshift.

SGR 1806-20 Superflare on 27 December 2004

Many spacecraft, including Swift, observed the giant flare from SGR 1806-20 that occurred on 27 December 2004. An SGR (Soft Gamma Repeater) is an extremely strongly magnetized neutron star, called a "magnetar". These are the most strongly magnetized objects in the Universe. "Starquakes" on such stars can cause reconnection within the magnetic field, which results in one or more bursts of soft gamma-rays.

The December 27 event was the brightest gamma-ray transient ever observed, even brighter than any GRB; for 0.2 seconds the total energy flux of the flare received at Earth was brighter than what is received from the full moon! Indeed, the burst was powerful enough to briefly illuminate the moon in X-rays. The burst lasted 380 seconds in total. The X-rays and gamma-rays from the burst briefly ionized Earth's atmosphere, and the effects of this were observed around the globe.

Hard X-Ray/Gamma-Ray Sky Survey

Swift is carrying out the most sensitive hard X-ray sky survey ever. The precise energy range is from 14 to 195 keV. Each five minutes that passes as Swift watches the sky for GRBs contributes one snapshot in 15 different energy bins to the X-ray sky survey. The first three months of the survey has a flux limit of roughly 10^{-11} erg cm^{-2} s^{-1}, with positions accurate to roughly 2.7 arc-minutes for the faintest sources. The final survey will have a similar flux limit over the entire sky.

The first reported catalog from this survey contains 66 objects, 56 of which have been identified, the majority of which are AGNs (Active Galactic Nuclei). An AGN is a supermassive black hole sitting near the center of a galaxy that is powered by large amounts of gas falling onto it. Two of the AGNs in the catalog are powerful radio sources called "blazars" with redshifts greater than 0.1. The median redshift of the radio-quiet AGNs is 0.012.

■ ***Above**: Two examples of burst afterglow flares seen in XRT observations: XRF 050406 (left), and GRB 050502B (right). The vertical axis is the count rate in counts per second. The horizontal axis is time since the burst trigger in seconds. Image courtesy David Burrows et al.*

The first Swift catalog is the modest beginning of a grand catalog that should contain more than 400 AGNs, and will be a true sample of the AGN population. For example, one of the results from the first catalog is that virtually all low luminosity AGNs are surrounded by absorbing and obscuring gas. It is now expected that roughly two out of three AGNs that Swift detects will be previously unrecognized nearby examples that are highly absorbed. Some of these objects will be near enough to resolve at other wavelengths, and such follow-up studies will improve our understanding of the AGN mechanism.

Furthermore, at the time of this writing, Swift has now detected six blazars with redshifts greater than 2, three of which were previously unknown. Blazars have strong emissions at gamma-ray energies, and these radiations will be redshifted to the X-ray region of the spectrum in very distant examples. Thus, Swift has opened the door to the study of the most distant members of this most powerful class of AGNs.

Type Ia supernova: SN 2005am

The Type Ia supernova named SN 2005am in the spiral galaxy NGC 2811 was observed with the Swift UVOT from 4 days before the peak in the "blue" light curve to 69 days after the peak. This supernova was quite widely observed by ground-based telescopes, but Swift provided extremely high quality and stable UBV (Ultraviolet-Blue-Visual) light curves for this very important class of cosmological distant indicator.

The Swift observations provide well-sampled light curves shorter than 2,500 Angstroms, which allow a unique study of the ultraviolet color evolution of a Type Ia supernova. This is important, because Type Ia supernovae at higher redshifts have this part of the spectrum redshifted into the more easily observable blue and visual bands. Therefore, the Swift results help with the interpretation of the observations of the more distant examples. The UVOT light curves also constrain theoretical models of Type Ia supernovae.

Spectral ('grism') observations were also carried out via the UVOT, and Swift also provided an upper limit to the supernova's X-ray luminosity via the XRT.

It should be noted, for comparison, that NGC 2811 lies at a redshift of only 0.0079, which is vastly smaller than that of even the nearest GRBs. This puts the awesome power of the bursts into perspective.

Very Unusual GRB of 18 February 2006

The Swift BAT detected a very unusual GRB on 18 February 2006 that turned out to be the precursor event of a supernova. The burst occurred in a galaxy at a redshift of 0.0331, which corresponds to a distance of only 440 million light years, to make it the second closest GRB ever recorded. The burst lasted 33 minutes, which is a hundred times longer than the typical GRB. The burst was intrinsically dim, had a very soft spectrum, and displayed a relatively smooth gamma-ray light curve. A radio afterglow was detected in record early time. The optical afterglow initially grew brighter for several days, and eventually displayed the spectrum of a Type Ib/c supernova. Named SN 2006aj, this supernova turns out to be the best observed supernova ever at early times, thanks to the quick detection via Swift, and the prompt and thorough follow up that is typically received by unusual GRBs. Time will tell how this very special event fits into the big picture.

The Future

Swift should continue to provide many more observations in the coming years of gamma-ray bursts, soft gamma repeaters, active galactic nuclei, and Type Ia supernovae, and, hence, Swift will continue to facilitate scientific advances in our understandings of these astrophysically extreme objects.

In particular, the GRBs should be detectable out to a redshift of at least 10, making them the most distant observable objects in the Universe, and, consequently, the bursts are probes of both the early Universe, and the space through which their radiations must travel to reach us. The GRBs discovered by Swift promise to provide unique data for studying the star formation history of the Universe, the epoch of reionization of the InterGalactic Medium (between when the first stars and first QSOs formed), the metallicity history of the Universe, and the dust and gas content of early galaxies.

The nominal mission lifetime of Swift is two years, but an extension can be anticipated if the instruments continue to function well, and scientifically useful results continue to be produced. The Swift spacecraft should remain in Earth orbit until at least 2022, some time after which atmospheric drag will cause it to re-enter Earth's atmosphere. We believe we have only witnessed the beginning of the great discoveries of the Swift era.

APPENDICES

The four unit telescopes of the ESO very Large Telescope (VLT) atop Cerro Paranal in Chile. Image courtesy European Southern Observatory.

APPENDIX 1 - CURRENT OPTICAL TELESCOPES

Aperture (Meters)	Telescope Name	Location	Web Link	Details
10.0	Keck	Mauna Kea Observatory, Hawaii. William Keck Observatory.	http://www.keckobservatory.org/	Segmented mirror telescope.
	Keck II		http://www.ifa.hawaii.edu/mko/	Can be used as an interferometer
10.0	SALT	South African Astronomical Observatory	http://www.salt.ac.za/	Spherical Mirror, fixed elevation
9.2	Hobby-Eberly	Mt. Fowlkes, Texas	http://www.as.utexas.edu/mcdonald/het/het.html	Spherical Mirror, fixed elevation
8.4	Large Binocular Telescope (LBT)	Mt. Graham, Arizona	http://medusa.as.arizona.edu/lbto/	Twin pair of 8.4-m mirrors with light gathering equivalent of an 11.8m and the resolution of a 23-m
			http://mgpc3.as.arizona.edu/	
8.3	Subaru	Mauna Kea, Hawaii	http://www.naoj.org/	National Astronomical Observatory of Japan
8.2	VLT - Antu	Cerro Paranal, Chile	http://www.eso.org/paranal/	The Very Large Telescope (VLT). Four 8.2 meter telscopes that can operate independently, or as a combined interferometer
	VLT - Kueyen			
	VLT - Melipal			
	VLT - Yepun			
8.1	Gillett	Mauna Kea, Hawaii	http://www.gemini.edu/	Gemini North
			http://www.noao.edu/usgp/usgp.html	
	Gemini South	Cerro Pachon, Chile		Gemini South
6.5	MMT	Mt. Hopkins, Arizona	http://cfa-www.harvard.edu/mmt/	Originally a multi-mirror telescope with 6 mirrors, now using a single mirror but the MMT name has been kept for historical reasons.
	Walter Baade	Las Campanas Observatory, La Serena, Chile	http://www.lco.cl/magellan/	Magellan I
	Landon Clay			Magellan II
6.0	Bolshoi Azimuthal Telescope	Nizhny Arkhyz, Russia	http://www.sao.ru/Doc-en/index.html	Large Altazimuth Telescope
	Large Zenith Telescope, LZT	British Columbia, Canada	http://www.astro.ubc.ca/LMT/lzt/index.html	A 6-m liquid mirror aimed at the zenith
5.0	Hale	Palomar Mountain, California	http://www.astro.caltech.edu/observatories/palomar/	Home of the famous 200" Hale telescope
4.2	William Herschel	La Palma, Canary Islands, Spain	http://www.ing.iac.es/PR/wht_info/	Observatorio del Roque de los Muchachos
	SOAR	Cerro Pachon, Chile	http://www.soartelescope.org	Brazil/USA; CTIO
4.0	Victor Blanco	Cerro Tololo, Chile	http://www.ctio.noao.edu/telescopes/4m/base4m.html	Cerro Tololo Inter-American Observatory

APPENDIX 1 - CURRENT OPTICAL TELESCOPES

Aperture (Meters)	Telescope Name	Location	Web Link	Details
3.9	Anglo-Australian Telescope (AAT)	Coonabarabran, NSW, Australia	http://www.aao.gov.au/	The AAT at Siding Spring Observatory
3.8	Mayall	Kitt Peak, Arizona	http://www.noao.edu/kpno/	Part of the National Optical Astronomy Observatories
	UK Infrared Telescope (UKIRT)	Mauna Kea, Hawaii	http://www.jach.hawaii.edu/UKIRT/	World's largest dedicated infrared telescope
3.7	Advanced Electro Optical System	Maui, Hawaii	http://www.maui.afmc.af.mil/	Largest telescope for tracking satellites, largely military applications, part of the Maui Space Surveillance System (MSSS).
3.6	"360"	Cerro La Silla, Chile	http://www.ls.eso.org/lasilla/Telescopes/360cat/html/tel360.html	European Southern Obs.
	Canada-France-Hawaii (CFHT)	Mauna Kea, Hawaii	http://www.cfht.hawaii.edu/	Optical and infrared telescope
			http://cdsweb.u-strasbg.fr:2001/	
	Telescopio Nazionale Galileo	La Palma, Canary Islands, Spain	http://www.tng.iac.es/	Operated by the Italian National Institute for Astrophysics (INAF)
3.5	MPI-CAHA	Calar Alto, Spain	http://www.mpia-hd.mpg.de/Public/index_en.html	Operated by the Max Plank Institute for Astrophysics
			http://www.caha.es/	
	New Technology Telescope (NTT)	Cerro La Silla, Chile	http://www.ls.eso.org/lasilla/sciops/ntt/index.html	European Southern Observatory
			http://www.eso.org/welcome.html	
	Astrophysical Research Consortium (ARC)	Apache Point, New Mexico	http://tycho.apo.nmsu.edu:81/35m_manual/	Mostly remote controlled
	WIYN	Kitt Peak, Arizona	http://www.noao.edu/wiyn/wiyn.html	Consortium of Wisconsin, Indiana, Yale, and NOAO
	Starfire	Kirtland AFB, New Mexico	http://www.de.afrl.af.mil/SOR/3_5m_telescope.htm	Starfire Optical Range, military.
3.0	Shane	Mount Hamilton, California	http://mthamilton.ucolick.org/public/tele_inst/3m/	Lick Observatory
			http://mthamilton.ucolick.org/	
	NASA IRTF	Mauna Kea, Hawaii	http://irtfweb.ifa.hawaii.edu/	Dedicated infrared telescope
2.7	Harlan J. Smith Telescope	Mt. Locke, Texas	http://www.as.utexas.edu/mcdonald/mcdonald.html	McDonald Observatory
2.6	BAO	Byurakan, Armenia	http://www.aras.am/bao.html	Byurakan Astrophysical Observatory
			http://www.bao.am/	
	Shajn	Crimea, Ukraine	http://www.crao.crimea.ua/craoinfo/stella.html	Crimean Astrophysical Observatory

APPENDIX 1 - CURRENT OPTICAL TELESCOPES

Aperture (Meters)	Telescope Name	Location	Web Link	Details
2.5	100" Hooker Telescope	Mt. Wilson, California	http://www.mtwilson.edu/ http://www.mtwilson.edu/vir/100/index.php	Famous 100" telescope used by Edwin Hubble to determine galactic distance scale.
	Isaac Newton	La Palma, Canary Islands, Spain	http://www.ing.iac.es/PR/int_info/	Observatorio del Roque de los Muchachos
	Nordic Optical		http://www.not.iac.es/	
	Irénée du Pont Telescope	La Serena, Chile	http://www.lco.cl/lco/dupont/	Las Campanas Observatory
	Sloan Digital Sky Survey	Apache Point, New Mexico	http://www.sdss.org/sdss.html	120 megapixel camera images 1.5 sq degrees. Huge survey instrument.
2.5	CHARA	Mt. Wilson, California	http://www.chara.gsu.edu/CHARA/array.html	Center for High Angular Resolution Astronomy - Interferometer array using six 1-m telescopes
2.4	Hiltner	Kitt Peak, Arizona	http://www.astro.lsa.umich.edu/obs/mdm/technical/hiltner.html	MDM Observatory, operated by a consortium of universities
	Hubble Space Telescope	Low Earth orbit	http://www.stsci.edu/ http://hubblesite.org/	Launched in 1990, The Hubble Space Telescope remains in low Earth orbit.

Above: *The Gemini North telescope. Image courtesy Gemini Observatory/AURA.*

Left: *The four unit telescopes of the ESO Very Large Telescope. Image courtesy European Southern Observatory.*

Far Left: *The two Keck domes atop the summit of Mauna Kea. Image courtesy W. M. Keck Observatory/CARA.*

APPENDIX 2 - FUTURE OPTICAL TELESCOPES

Aperture (meters)	Name	Location	Proposed Completion Date	Web Link	Comments
100	Overwhelmingly Large Telescope (OWL)	undecided	2020+	http://www.eso.org/projects/owl/	A huge segmented mirror that would be built over a long period - gradually adding new mirror segments in a production line process leading to a 100-m telescope with milliarcsecond resolution.
30-60	European Extremely Large Telescope (E-ELT)	undecided	Baseline design 2006	http://www.eso.org/projects/e-elt/	Developed as a more modest design than the 100-m OWL and achievable in a shorter timescale and as a technology test for the larger instrument.
30-50	Giant Segmented Mirror Telescope (GSMT)	undecided	2012	http://www.noao.edu/future/gsmt.html	Concept study for a 30-50-m telescope performed in 2001.
50	Euro50	Chile or La Palma, Canary Islands	2014	http://www.astro.lu.se/%7Etorben/euro50/	Consortium of institutes from Finland, Ireland, Spain, Sweden and United Kingdom to build 618 2-m mirror segments forming a 50-m telescope.
30	Thirty Meter Telescope (TMT)	Undecided	2015	http://www.tmt.org/	More than 700 hexagonal elements spanning 30-meters, site seletion 2007 and proposed completion 2015. The project is a combination of three precursor projects.
21.4 (7x8.4)	Giant Magellan Telescope (GMT)	Chile?	2016	http://www.gmto.org/	Six off-axis segments and one central segment forming one huge mirror.
10.4	Gran Telescopio Canarias	La Palma, Canary Islands, Spain	2006	http://www.gtc.iac.es/home.html	Segmented mirror similar to the Keck Telescopes.
8	Large Synoptic Survey Telescope	Chile	2012	http://www.lsst.org/lsst_home.shtml	A fast wide field survey telescope.
4.2	LAMOST	Xinglong Station, China	2007	http://www.lamost.org/en/	Large Sky Area Multi-Object Fiber Spectroscopic Telescope.
	Discovery Channel Telescope	Happy Jack, Arizona	2009	http://www.lowell.edu/DCT/index.html	Discovery Channel and other partners.
4	Visible and Infrared Survey Telescope	Cerro Paranal, Chile	2007	http://www.vista.ac.uk/	Wide field survey telescope for visible and infrared astronomy.

APPENDIX 2 - FUTURE OPTICAL TELESCOPES

Aperture (meters)	Name	Location	Proposed Completion Date	Web Link	Comments
2.5	SOFIA	Stratosphere	2006	http://www.sofia.usra.edu/	Stratospheric Observatory for Infrared Astronomy - on board a modified Boeing 747 to be based at Ames Research Center.
2.3	Aristarchos	Mt. Helmos, Greece	2006	http://www.astro.noa.gr/ASC_2.3m/ngt_main.htm	Under final testing in Greece.
4x1.8	Pan-STARRS	Hawaii	Prototype 2006	http://pan-starrs.ifa.hawaii.edu/public/	Four individual telescopes on one mount, a precursor to LSST. Very large digital cameras will image 6,000^2 deg per night in survey mode.
2.4+10x1.4	Magdalena Ridge Observatory	Socorro, New Mexico	2008/9	http://www.mro.nmt.edu/	A single 2.4-m telescope and ten 1.4-m interferometer.

FUTURE COSMIC RAY DETECTORS

3000 km^2	Pierre Auger	Argentina	2006	http://www.auger.org/	Cerenkov detectors, partially operational in 2005
4 x 12-m	Veritas	Whipple Observatory, Arizona	2006	http://veritas.sao.arizona.edu/	First Veritas telescope operational February 2005.

APPENDIX 3 - THE LARGEST NON-OPTICAL TELESCOPES

SUB-MILLIMETER AND RADIO TELESCOPES

Aperture (Meters)	Number of Antennae	Baseline	Telescope Name	Location	Web Link	Details
576.0	895		RATAN-600	Russia	http://www.sao.ru/ratan/index.html.en	An array of 11m x 2m panels in a circle, 576 m across.
305.0	1		Arecibo Radio Telescope	Puerto Rico.	http://www.naic.edu/	The largest single dish radio telescope in the world.
100 x 110	1		Robert C Byrd Green Bank Telescope	Green Bank, West Virginia	http://www.gb.nrao.edu/	The largest steerable single dish in the world.
100.0	1		Effelsberg Radio Telescope	Effelsberg, Germany	http://www.mpifr-bonn.mpg.de/index_e.html	Second largest steerable single dish in the world
76.0	1		The Lovell Telescope	Jodrell Bank, University of Manchester, England	http://www.jb.man.ac.uk/	
64.0	1		Parkes Radio Telescope	Parkes, New South Wales, Australia	http://www.parkes.atnf.csiro.au/	Largest southern hemisphere radio telescope, famous for relaying Apollo 11 signals and the movie, The Dish, in addition to cutting edge research.
40.0	1		Owens Valley Radio Observatory	California	http://www.ovro.caltech.edu/	
37.0	1		Haystack Observatory	Massachussetts	http://www.haystack.edu/haystack/haystack.html	
30.0	1		Institute of Millimeter Radio Astronomy	Granada, Spain	http://iram.fr/	
26.0	1		Penticton Radio Telescope	Dominion Astrophysical Observatory	http://www.drao-ofr.hia-iha.nrc-cnrc.gc.ca/facilities/telescopes/26m/	
26.0	1		Hartebeesthoek Observatory	South Africa	http://www.hartrao.ac.za/summary/sumeng.html	Originally one of NASA's Deep Space tracking dishes.
26.0	1		Mount Pleasant	Tasmania	http://www-ra.phys.utas.edu.au/observatories/mount-pleasant.html	
25.0	27	36 km	Very Large Array (VLA)	Socorro, New Mexico	http://www.vla.nrao.edu/	Antenna arranged in a large 'Y' shape.
25.0	10	>3,000 km	Very Long Baseline Array (VLBA)	Across the United States	http://www.vlba.nrao.edu/	10 identical antennae located across the USA from Hawaii to the U.S. Virgin Islands
25.0	14	2.7 km	Westerbork Synthesis Radio Telescope	Holland	http://www.astron.nl/	
15.0	6	408-m	IRAM Plateau de Bure Interferometer array	Granada, Spain	http://iram.fr/IRAMES/index.htm	
15.0	1		James Clerk Maxwell Telescope (JCMT)	Mauna Kea, Hawaii	http://www.jach.hawaii.edu/JCMT/	JCMT is the largest sub-millimetre radio telescope in the world, and is operated by the UK, Canada and Holland.
13.0	5	4.8 km	The Ryle Telescope	Cambridge, England	http://www.mrao.cam.ac.uk/telescopes/ryle/	Mullard Radio Astronomy Observatory, part of the Cavendish Laboratory, University of Cambridge.

APPENDIX 3 - THE LARGEST NON-OPTICAL TELESCOPES

SUB-MILLIMETER AND RADIO TELESCOPES

Aperture (Meters)	Number of Antennae	Baseline	Telescope Name	Location	Web Link	Details
12.0	1		Arizona Radio Observatory		http://kp12m.as.arizona.edu/	Arizona Radio Observatory/Steward Observatory/University of Arizona 12m telescope, Tucson, USA.
10.4	1		Leighton submillimeter telescope	Mauna Kea, Hawaii.	http://www.submm.caltech.edu/cso/	Caltech Submillimeter Observatory
10.4 & 6.1	15	2 km	Combined Array for Research in Millimeter-wave Astronomy (CARMA)	California	http://www.mmarray.org/	Combination of Owens Valley Radio Telescope and the Berkeley-Illinois-Maryland Association observatory to create a new large millimeter array.
9.0	7	600 m	The Synthesis Telescope	Dominion Astrophysical Observatory	http://www.drao-ofr.hia-iha.nrc-cnrc.gc.ca/	Aperture synthesis telescope
778 x 12	2		Molonglo Observatory Synthesis Telescope	Canberra, Australia	http://www.physics.usyd.edu.au/astrop/most/	Two cylindrical paraboloid antennas 15-m apart.
Yagi	60	4.6 km	Cambridge Low Frequency Synthesis Telescope	Cambridge, England	http://www.mrao.cam.ac.uk/telescopes/clfst/	Performing the 7C survey of the northern hemisphere
Various	6	217 km	MERLIN VLBI array	Operated by Jodrell Bank	http://www.merlin.ac.uk/	Multi-Element Radio-Linked Interferometer Network.

APPENDIX 3B - FUTURE NON-OPTICAL TELESCOPES

Aperture (Meters)	Number of Antennae	Baseline	Proposed Completion Date	Telescope Name	Location	Web Link	Details
12.0	64	Up to 18 km	2012	Atacama Large Millimeter Array	Llano de Chajnantor, Chile	http://www.mma.nrao.edu/	Proposed 64 antenna sub-millimeter array located in the Chilean Andes. Review underway to reduce number of antennae to 40 or 50, due to US budget restrictions.
30-300	Up to 1,000	1 Sq km	2010-2020	Square Kilometer Array (SKA)	South Africa	http://www.ska.ac.za/	Phase 1 complete in 2010 with full array by 2020, providing extremely high angular resolution.
50.0	1	-	2008	Karoo Array Telescope	South Africa	http://www.ska.ac.za/kat/project.html	A demonstrator for the SKA technology, located within 50 km of the proposed SKA site.
50.0	1	-	2008	Large Millimeter Telescope/Gran Telescopio Millimetrico	Volcán Sierra Negra, Central Mexico	http://www.lmtgtm.org/	Collaborative project between USA and Mexico, to produce the largest millimeter telescope in the world.

APPENDIX 4 - OPTICAL INTERFEROMETERS

Name	Location	Array Elements (#, mirror size)	Maximum baseline (meters)	Web Links	Notes
COAST	Cambridge, England	5 (40 cm)	67	http://www.mrao.cam.ac.uk/telescopes/coast/	Operational since 1991
GI2T/REGAIN	Cote D'Azure Observatory	2 (1.5 m)	65	http://www.obs-nice.fr/	Operational since 1985
InfraRed Spatial Int. (ISI)	Mt. Wilson, California	3 (1.65)	75	http://isi.ssl.berkeley.edu/isi.html	Operational since 1990
SUSI (Sydney University)	Narrabri, Australia	13 (20 cm)	640	http://www.physics.usyd.edu.au/astron/susi/	Operational since 1991
IOTA (Harvard - CfA)	Mt. Hopkins	3 (45 cm)	38	http://www.cfa.harvard.edu/cfa/oir/IOTA/	Closure due July 1, 2006
NPOI (USNO)	Anderson Mesa, Arizona	6 (50 cm)	437	http://www.nofs.navy.mil/projects/npoi/	Web site under revision
Palomar Testbed Interferometer	Palomar Mountain, California	3 (40 cm)	110	http://pti.jpl.nasa.gov/	Operational since 1995
CHARA Array (Georgia St.U.)	Mt. Wilson, California	6 (1.0 m)	330	http://joy.chara.gsu.edu/CHARA/index.html	Operational since 1999
Keck Interferometer	Mauna Kea, Hawaii	2 (10 m)	85	http://planetquest.jpl.nasa.gov/Keck/keck_index.cfm	Operational since 2001
VLT Interferometer (ESO)	Cerro Paranal	4 (8.5 m) and 4 (1.8 m)	200 (140)	http://www.eso.org/projects/vlti/	Operational since 2001
MIRA-1.2 (University of Tokyo)	Mitaka, Japan	2 (30 cm)	30	http://tamago.mtk.nao.ac.jp/mira/index_uk.html	Operational since 2002
LBT Interferomter	Mt. Graham, Arizona	2 (8.4 m)	22.8	http://lbti.as.arizona.edu/	Operational since 2006
Magdelena Ridge Observatory	Mt. Baldy, New Mexcio	10 (1.5 m)	400	http://www.mro.nmt.edu/Interferometer/index.php	Under construction, completion due 2008/09

CONCEPTS UNDER STUDY

Name	Location	Array Elements (#, mirror size)	Maximum baseline (meters)	Web Links	Notes
SIM Planetquest	Proposed in-orbit instrument			http://planetquest.jpl.nasa.gov/SIM/sim_index.cfm	
Terestrial Planetfinder	Proposed in-orbit instrument			http://planetquest.jpl.nasa.gov/TPF/tpf_index.cfm	
DARWIN (ESA)	Proposed in-orbit instrument	3+ (3+ m)		http://sci.esa.int/darwin	2015
Antarctic Plateau Interferometer	Antarctica	3+ (2 to 4 m)		http://www-laog.obs.ujf-grenoble.fr/heberges/API/	
Optical Very Large Array	France	-	-	http://www.oamp.fr/lise/projets.html	
Stellar Interferometer	Proposed in-orbit instrument	20-30 (1 m)	100 to 1000	http://hires.gsfc.nasa.gov/si/	
Large Aperture Mirror Array (LAMA)	Potential sites in Chile and New Mexico	18 (10 m)	54	http://www.astro.ubc.ca/lmt/lama/overview.html	Proposes liquid mirror telescopes
Micro Arcsecond X-Ray Imaging (MAXIM)	Proposed in-orbit instrument			http://maxim.gsfc.nasa.gov/	
SPECS	Proposed in-orbit instrument	3 (3 m)		http://space.gsfc.nasa.gov/astro/specs/	2010+
OSIRIS (Russia)	Proposed in-orbit instrument	-	-	http://www.inasan.rssi.ru/eng/osiris/index.html	2018
Optical Hawaiian Array for nanoradian Astronomy	Mauna Kea, Hawaii	6 (3 to 10 m)	800	http://www.cfht.hawaii.edu/~lai/ohana.html	Under development, completion due 2008/09

Note: A useful web reference is the Optical Long Baseline Interferometry News web site maintained at the Jet Propulsion Laboratory, California. http://olbin.jpl.nasa.gov/

APPENDIX 5.1 - SPACE OBSERVATORIES

Space Observatories	Launch Date	Type	Web link	Details
Suzaku (Astro-E2)	9-Jul-05	X-ray	http://www.isas.jaxa.jp/e/enterp/missions/suzaku/index.shtml	Japanese X-ray astronomy satellite.
MOST	30-Jun-03	Optical	http://www.astro.ubc.ca/MOST/	Canadian space telescope searching for microvariability of stars.
GALEX	28-Apr-03	UV	http://www.galex.caltech.edu/	Ultraviolet observatory studying star formation regions in distant galaxies.
FUSE	24-Jun-99	UV	http://fuse.pha.jhu.edu/	Far Ultraviolet Spectroscopic Explorer.
Spitzer	25-Aug-03	IR	http://www.spitzer.caltech.edu/	Spitzer Infrared Space Telescope.
Hubble	25-Apr-90	Optical, UV, IR	http://hubblesite.org/	Hubble Space Telescope.
Rossi X-ray Timing Explorer	30-Dec-95	X-ray	http://heasarc.gsfc.nasa.gov/docs/xte/xtegof.html	Studying black holes, neutron stars, x-ray pulsars.
XMM-Newton	10-Dec-99	X-ray	http://xmm.vilspa.esa.es/	Europe's X-ray nested mirror telescope.
Integral	17-Oct-02	Gamma ray	http://sci.esa.int/science-e/www/area/index.cfm?fareaid=21	Europe's gamma ray observatory.
Akari (Astro-F)	23-Feb-06	IR	http://www.jaxa.jp/missions/projects/sat/astronomy/astro_f/index_e.html	Japanese infrared imaging observatory.
SWIFT	20-Nov-04	Gamma ray	http://www.nasa.gov/mission_pages/swift/main/index.html	Gamma Ray Observatory studying gamma ray bursts.
Chandra	3-Jul-99	X-ray	http://chandra.harvard.edu/index.html	Large X-ray nested mirror observatory.
Gravity Probe-B	20-Apr-04	Gyroscopes	http://einstein.stanford.edu/	Relativity gyroscope experiment collected data from August 2004 to September 2005.
HETE-2	9-Oct-00	Gamma ray	http://space.mit.edu/HETE/spacecraft.html	High Energy Transient Explorer-2 multiwavelength gamma ray observatory.
WMAP	30-Jun-01	IR	http://map.gsfc.nasa.gov/	Wilkinson Microwave Anisotropy Probe.

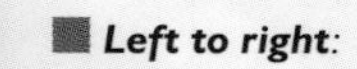

***Left to right*:**

Artist's impression of the Chandra X-ray Observatory. Image courtesy Chandra X-ray Center and NASA.

The Hubble Space Telescope in orbit. Image courtesy NASA/Space Telescope Science Institute.

Artist's impression of the Spitzer Infrared Space Observatory. Image courtesy NASA/ Jet Propulsion Laboratory - Caltech.

APPENDIX 5.2 - FUTURE SPACE OBSERVATORIES

Space Observatories	Projected Launch Date	Type	Web link	Details
Constellation-X	2011	X-ray	http://constellation.gsfc.nasa.gov/	Array of X-ray telescopes flying in formation and operating together to simulate one large X-ray telescope.
COROT	2006	Optical	http://sci.esa.int/science-e/www/area/index.cfm?fareaid=39	European mission to search for rocky planets around other stars by observing transits.
Darwin	2015	Optical	http://sci.esa.int/science-e/www/area/index.cfm?fareaid=28	Three 3-m telescopes searching for Earth-like planets.
GAIA	2011	Astrometry	http://sci.esa.int/science-e/www/area/index.cfm?fareaid=26	Precision astrometry on a billion stars in our galaxy and local group.
GLAST	2007	Gamma ray	http://www-glast.sonoma.edu/	Gamma Ray Large Area Space Telescope follow on mission to SWIFT
Herschel (previously FIRST)	2007	IR and Sub-mm	http://sci.esa.int/science-e/www/area/index.cfm?fareaid=16	First space telescope with a deployable mirror (3.5-m) to cover far IR and sub-mm.
Hyper	2020		http://sci.esa.int/science-e/www/area/index.cfm?fareaid=46	Testing relativity and fundamental forces of gravity and electromagnetism.
James Webb Space Telescope	2013	IR	http://www.jwst.nasa.gov/	Deployable 6.5-meter space telescope - the replacement for the Hubble Space Telescope.
Kepler	2008	Optical	http://www.kepler.arc.nasa.gov/	Searching for transiting planets around nearby stars.
LISA	2015	Gravitational wave interferometer	http://lisa.jpl.nasa.gov/	Three spacecraft searching for gravitational waves - a space-based version of LIGO.
LISA-Pathfinder	2009	Gravitational wave detector	http://sci.esa.int/science-e/www/area/index.cfm?fareaid=46	European mission to test concepts for the LISA mission.
MAXIM	Proposed mission	X-ray	http://maxim.gsfc.nasa.gov/	Micro-arcsecond X-ray imaging Mission
OSIRIS (Russia)	Proposed mission	Astrometry	http://www.inasan.rssi.ru/eng/osiris/index.html	Russian proposed Optical Stellar Interferometer measuring precise stellar positions.
Planck	2007	Microwave	http://sci.esa.int/science-e/www/area/index.cfm?fareaid=17	Cosmic microwave background anisotropy survey.
Stellar Interferometer	2025	Interferometer	http://hires.gsfc.nasa.gov/si/	20-30 1-m space telescopes in variable formation, testbed mission proposed in 2015.
SIM PlanetQuest	2015	UV and optical Interferometer	http://planetquest.jpl.nasa.gov/SIM/sim_index.cfm	Space Interferometry Mission to search for planets around other stars.
SPECS	Proposed mission	IR	http://space.gsfc.nasa.gov/astro/specs/	Proposed sub-millimeter probe of Cosmic Sctructure.
Stratospheric Observatory for Infrared Astronomy (SOFIA)	2006	IR	http://www.sofia.usra.edu/	2.5-m telescope onboard modified 747-SP aircraft, based at Ames Research Center, California.
WISE	2009	IR	http://wise.ssl.berkeley.edu/	Wide-field Infrared Survey Explorer, all-sky survey to complement the JWST.
XEUS	2015	X-ray	http://sci.esa.int/science-e/www/area/index.cfm?fareaid=25	European next generation X-ray telescope.

APPENDIX 6

EXTRASOLAR PLANETS

The growing list of extrasolar planets detected by a variety of methods is displayed in the following table. A variety of teams and a range of methods are used to detect extrasolar planets. Most common is the radial velocity method, using extremely high precision spectroscopy to measure the back and forth motion of a star induced by the gravitational interaction with an orbiting planet. This method is most sensitive to the largest of any planets orbiting close to the parent star. Longer orbital periods and smaller mass planets require longer time periods of observations and higher precision to detect.

A method that is beginning to show results is through detection of gravitational microlensing. Starlight from a very distant star is momentarily intensified by the passage of a nearby star directly in its line of sight. The gravitational attraction of the nearby star acts as a lens, focusing light following principles laid down by Albert Einstein in his General Theory of Relativity. Secondary 'blips' in the light curve can occur if a planet also happens to be orbiting the star.

A third method is using the transit technique. A planet passes directly in front of a star as seen from Earth causing a brief diminution in the stars brightness.

Finally, the surprising discovery of planets orbiting pulsars is based upon detection of the sinusoidal variation the arrival of pulses at the Earth. A massive planet will cause the pulsar to rock back and forth around the mutual center of gravity of the system. Instead of using variations in spectral Doppler shift of spectral lines, the pulse timings are affected.

Each month new discoveries are announced. You can keep track of these new discoveries via the PlanetQuest web site hosted at the Jet Propulsion Laboratory in Pasadena, California. The following table is based upon their data.

Go to
http://planetquest1.jpl.nasa.gov/atlas/atlas_index.cfm
for the latest updates.

APPENDIX 6 - EXTRASOLAR PLANETS

Planet	Star	Constellation	Year of Discovery	Discovered By	Orbital Radius	Orbit Period (seconds)	Planet Type	Planet Mass	Orbit Eccentricity	Method of Detection	Star Distance from Sun (Light Years)
PSR 1257 a	PSR 1257		1991	Alexander Wolszczan and Dale Frail, National Radio Astronomy Observatory	0.19 AU	25.262s	Pulsar	0 (Earth = 1)	-1	Timing	978
PSR 1257 b	PSR 1257		1991	Alexander Wolszczan and Dale Frail of the National Radio Observatory in New Mexico	0.36 AU	66.5s	Pulsar	0 (Earth = 1)	0.1	Timing	978
PSR 1257 c	PSR 1257		1994		0.46 AU	98.2s	Pulsar	0 (Earth = 1)	0.02	Timing	978
PSR 1257 d	PSR 1257		1994		40 AU	62050s	Pulsar	0 (Earth = 1)	-1	Timing	978
51 Pegasi b	51 Pegasi	Pegasus	1995	Michel Mayor and Didier Queloz, Geneva Observatory	0.05 AU	4.23s	Hot Jupiter	0.47 (Jupiter = 1)	0	Radial Velocity (or Doppler Spectroscopy)	48
Upsilon Andromedae b	Upsilon Andromedae	Andromedae	1996	Geoffrey Marcy and R. Paul Butler, San Francisco State University	0.05 AU	4.62s	Hot Jupiter	0.71 (Jupiter = 1)	0.034	Radial Velocity (or Doppler Spectroscopy)	43.9
55 Cancri b	55 Cancri	Cancer	1996	San Francisco State University Planet Search	0.118 AU	14.66s	Gas Giant	0.84 (Jupiter = 1)	0.03	Radial Velocity (or Doppler Spectroscopy)	44
47 Ursae Majoris b	47 Ursae Majoris	Ursa Major	1996	University of California Planet Search	2.1 AU	1095s	Gas Giant	2.41 (Jupiter = 1)	0.096	Radial Velocity (or Doppler Spectroscopy)	43
tau Boo	tau Bootis	Bootes	1996	San Francisco State University Planet Search	0.0462 AU	3.3128s	Hot Jupiter	3.87 (Jupiter = 1)	0.018	Radial Velocity (or Doppler Spectroscopy)	49
70 Virginis b	70 Virginis	Virgo	1996	San Francisco State Planet Search	0.43 AU	116.6s	Gas Giant	6.6 (Jupiter = 1)	0.4	Radial Velocity (or Doppler Spectroscopy)	59
rho CrB	rho Coronae Borealis	Corona Borealis	1997	Smithsonian Astrophysical Observatory; National Center for Atmospheric Research; Penn State U.	0.23 AU	39.645s	Gas Giant	1.1 (Jupiter = 1)	0.028	Radial Velocity (or Doppler Spectroscopy)	55
16 Cygni b	16 Cygni	Cygnus	1997	San Francisco State University Planet Search	1.7 AU	804s	Gas Giant	1.5 (Jupiter = 1)	0.67	Radial Velocity (or Doppler Spectroscopy)	70
HD 217107 b	HD 217107	Pisces	1998	Keck and Lick Observatories	0.07 AU	7.11s	Hot Jupiter	1.28 (Jupiter = 1)	0.14	Radial Velocity (or Doppler Spectroscopy)	121
HD 210277 b	HD 210277	Aquarius	1998	Keck Observatory	1.097 AU	437s	Gas Giant	1.28 (Jupiter = 1)	0.45	Radial Velocity (or Doppler Spectroscopy)	72

APPENDIX 6 - EXTRASOLAR PLANETS

Planet	Star	Constellation	Year of Discovery	Discovered By	Orbital Radius	Orbit Period (seconds)	Planet Type	Planet Mass	Orbit Eccentricity	Method of Detection	Star Distance from Sun (Light Years)
HD 187123 b	HD 187123	Cygnus	1998	Keck Observatory	0.042 AU	3.1s	Hot Jupiter	0.52 (Jupiter = 1)	0.03	Radial Velocity (or Doppler Spectroscopy)	153
Gliese 876 b	Gliese 876	Aquarius	1998	1998, Haute-Provence Observatory and SFSU Planet Search	0.21 AU	61.02s	Gas Giant	> 1.89 (Jupiter = 1)	0.27	Radial Velocity (or Doppler Spectroscopy)	15
HD 195019	HD 195019	Delphinus	1998	Keck and Lick observatories	0.14 AU	18.3s	Gas Giant	3.43 (Jupiter = 1)	0.05	Radial Velocity (or Doppler Spectroscopy)	65
HD 168443 b	HD 168443	Serpens	1998	Keck and Lick observatories	0.29 AU	57.9s	Gas Giant	7.2 (Jupiter = 1)	0.54	Radial Velocity (or Doppler Spectroscopy)	107
HD 168443 c	HD 168443	Serpens	1998	Keck and Lick observatories	2.87 AU	2135s	Gas Giant	17.1 (Jupiter = 1)	0.2	Radial Velocity (or Doppler Spectroscopy)	107
14 Herculis b	14 Herculis		1998	Haute-Provence Observatory	2.5 AU	1650s	Gas Giant	3.3 (Jupiter = 1)	0.326	Radial Velocity (or Doppler Spectroscopy)	55.42
HD 209458 b	HD 209458	Pegasus	1999	San Francisco State University Planet Search	0.045 AU	3.52s	Hot Jupiter	0.69 (Jupiter = 1)	0	Transit Method	153
HD 192263 b	HD 192263	Aquila	1999	Keck Precision Velocity Survey	0.15 AU	24.4s	Hot Jupiter	0.72 (Jupiter = 1)	0	Radial Velocity (or Doppler Spectroscopy)	65
HD 37124 b	HD 37124	Taurus	1999	Keck Precision Velocity Survey	0.585 AU	155s	Gas Giant	1.04 (Jupiter = 1)	0.19	Radial Velocity (or Doppler Spectroscopy)	107
HD 130322 b	HD 130322	Virgo	1999	Geneva Observatory	0.088 AU	10.724s	Gas Giant	1.08 (Jupiter = 1)	0.048	Radial Velocity (or Doppler Spectroscopy)	98
HD 177830 b	HD 177830	Taurus	1999	Keck Precision Velocity Survey	1 AU	391s	Gas Giant	1.28 (Jupiter = 1)	0.43	Radial Velocity (or Doppler Spectroscopy)	192
HD 134987 b	HD 134987	Libra	1999	Keck Precision Velocity Survey	0.78 AU	260s	Gas Giant	1.58 (Jupiter = 1)	0.25	Radial Velocity (or Doppler Spectroscopy)	81
HR 810 b	HR 810	Horologium	1999	ESO La Silla observatory	0.925 AU	320.1s	Gas Giant	2.26 (Jupiter = 1)	0.161	Radial Velocity (or Doppler Spectroscopy)	50.5

APPENDIX 6 - EXTRASOLAR PLANETS

Planet	Star	Constellation	Year of Discovery	Discovered By	Orbital Radius	Orbit Period (seconds)	Planet Type	Planet Mass	Orbit Eccentricity	Method of Detection	Star Distance from Sun (Light Years)
Upsilon Andromedae c	Upsilon Andromedae	Andromeda	1999	San Francisco State University and Smithsonian Center for Astrophysics	0.83 AU	241.2s	Gas Giant	2.11 (Jupiter = 1)	0.18	Radial Velocity (or Doppler Spectroscopy)	43.9
Upsilon Andromedae d	Upsilon Andromedae	Andromeda	1999	San Francisco State University and Smithsonian Center for Astrophysics	2.5 AU	1266.6s	Gas Giant	4.61 (Jupiter = 1)	0.41	Radial Velocity (or Doppler Spectroscopy)	43.9
HD 222582 b	HD 222582	Aquarius	1999	Keck Observatory	1.35 AU	576s	Gas Giant	5.4 (Jupiter = 1)	0.71	Radial Velocity (or Doppler Spectroscopy)	137
HD 10697 b	HD 10697	Andromeda	1999	Keck Precision Velocity Survey	2 AU	1083s	Gas Giant	6.59 (Jupiter = 1)	0.12	Radial Velocity (or Doppler Spectroscopy)	97.8
HD 83443 b	HD 83443	Vela	2000	Geneva Observatory	0.038 AU	2.99s	Hot Jupiter	0.35 (Jupiter = 1)	0.08	Radial Velocity (or Doppler Spectroscopy)	142
HD 168746 b	HD 168746	Scutum	2000	Geneva Observatory	0.066 AU	6.409s	Hot Jupiter	0.24 (Jupiter = 1)	0	Radial Velocity (or Doppler Spectroscopy)	141
HD 46375 b	HD 46375	Monoceros	2000	Marcy G., Butler P., Vogt S.	0.041 AU	3.024s	Hot Jupiter	0.249 (Jupiter = 1)	0	Radial Velocity (or Doppler Spectroscopy)	109
HD 108147 b	HD 108147	Crux	2000	Geneva southern extrasolar planet search program	0.098 AU	10.9s	Gas Giant	0.34 (Jupiter = 1)	0.558	Radial Velocity (or Doppler Spectroscopy)	126
HD 75289 b	HD 75289	Vela	2000	CORALIE survey for southern extrasolar planets	0.046 AU	3.51s	Hot Jupiter	0.42 (Jupiter = 1)	0.054	Radial Velocity (or Doppler Spectroscopy)	94.3
BD -10 3166 b	BD -10 3166	Crater	2000	California Planet Search Team	0.046 AU	3.487s	Hot Jupiter	0.48 (Jupiter = 1)	0	Radial Velocity (or Doppler Spectroscopy)	
HD 6434 b	HD 6434	Phoenix	2000		0.015 AU	22.09s	Gas Giant	0.48 (Jupiter = 1)	0.3	Radial Velocity (or Doppler Spectroscopy)	131
Epsilon Eridani b	Epsilon Eridani	Eridanus	2000	McDonald Observatory, with additional data from other planet searches	3.3 AU	2502.1s	Gas Giant	0.86 (Jupiter = 1)	0.608	Radial Velocity (or Doppler Spectroscopy), Astronomy	10.4

APPENDIX 6 - EXTRASOLAR PLANETS

Planet	Star	Constellation	Year of Discovery	Discovered By	Orbital Radius	Orbit Period (seconds)	Planet Type	Planet Mass	Orbit Eccentricity	Method of Detection	Star Distance from Sun (Light Years)
HD 38529 b	HD 38529	Orion	2000	San Francisco State University planet search	0.1293 AU	14.41s	Gas Giant	0.77 (Jupiter = 1)	0.28	Radial Velocity (or Doppler Spectroscopy)	138
HD 179949 b	HD 179949	Sagittarius	2000	Anglo-Australian Planet Search	0.045 AU	3.093s	Hot Jupiter	0.84 (Jupiter = 1)	0.05	Radial Velocity (or Doppler Spectroscopy)	88
HD 82943 b	HD 82943	Hydra	2000	CORALIE survey for southern extrasolar planets	1.16 AU	444.6s	Gas Giant	1.63 (Jupiter = 1)	0.41	Radial Velocity (or Doppler Spectroscopy)	89
HD 121504 b	HD 121504	Centaurus	2000	CORALIE survey for southern extrasolar planets	0.32 AU	64.6s	Gas Giant	0.89 (Jupiter = 1)	0.13	Radial Velocity (or Doppler Spectroscopy)	145
HD 52265 b	HD 52265	Monoceros	2000	Geneva Observatory	0.49 AU	118.96s	Gas Giant	1.13 (Jupiter = 1)	0.29	Radial Velocity (or Doppler Spectroscopy)	91
HD 27442 b	HD 27442	Reticulum	2000	Anglo-Australian Planet Search	1.18 AU	423s	Gas Giant	1.43 (Jupiter = 1)	0.02	Radial Velocity (or Doppler Spectroscopy)	59
HD 160691 b	HD 160691	Ara	2000	Anglo-Australian Planet Search	1.65 AU	743s	Gas Giant	1.97 (Jupiter = 1)	0.62	Radial Velocity (or Doppler Spectroscopy)	49
HD 19994 b	HD 19994	Cetus	2000	Geneva Observatory	1.3 AU	454s	Gas Giant	2 (Jupiter = 1)	0.2	Radial Velocity (or Doppler Spectroscopy)	73
HD 92788 b	HD 92788	Sextans	2000	San Francisco State University Planet Search	0.94 AU	340s	Gas Giant	3.8 (Jupiter = 1)	0.36	Radial Velocity (or Doppler Spectroscopy)	107
HD 12661 b	HD 12661	Aries	2000	San Francisco State University Planet Search	0.789 AU	264.5s	Gas Giant	2.83 (Jupiter = 1)	0.33	Radial Velocity (or Doppler Spectroscopy)	121
HD 169830 b	HD 169830	Sagittarius	2000	ESO La Silla Observatory	0.823 AU	230.4s	Gas Giant	2.96 (Jupiter = 1)	0.34	Radial Velocity (or Doppler Spectroscopy)	118
GJ 3021 b	GJ 3021	Hydrus	2000		0.49 AU	133.82s	Gas Giant	3.31 (Jupiter = 1)	0.505	Radial Velocity (or Doppler Spectroscopy)	57

APPENDIX 6 - EXTRASOLAR PLANETS

Planet	Star	Constellation	Year of Discovery	Discovered By	Orbital Radius	Orbit Period (seconds)	Planet Type	Planet Mass	Orbit Eccentricity	Method of Detection	Star Distance from Sun (Light Years)
Gliese 86 b	Gliese 86	Eridanis	2000	La Silla Observatory	0.11 AU	15.78s	Gas Giant	4 (Jupiter = 1)	0.046	Radial Velocity (or Doppler Spectroscopy)	36
HD 190228 b	HD 190228	Vulpecula	2000	Geneva Observatory	2.31 AU	1127s	Gas Giant	4.99 (Jupiter = 1)	0.43	Radial Velocity (or Doppler Spectroscopy)	215
HD 89744 b	HD 89744	Ursa Major	2000	Advanced Fiber-Optic Echelle (AFOE) spectrometer	0.88 AU	256s	Gas Giant	7.2 (Jupiter = 1)	0.7	Radial Velocity (or Doppler Spectroscopy)	130
HD 16141 b	HD 16141	Cetus	2000	Marcy G., Butler P., Vogt S., Keck Observatory	0.35 AU	75.8s	Gas Giant	0.215 (Jupiter = 1)	0.28	Radial Velocity (or Doppler Spectroscopy)	117
HD 162020 b	HD 162020		2000	ESO La Silla Observatory	0.072 AU	8.4s	Gas Giant	13.75 (Jupiter = 1)	0.277	Radial Velocity (or Doppler Spectroscopy)	101.9
HD 4208 b	HD 4208	Sculptor	2001	Keck Precision Velocity Survey	1.69 AU	829s	Gas Giant	0.81 (Jupiter = 1)	0.04	Radial Velocity (or Doppler Spectroscopy)	110
HD 82943 c	HD 82943	Hydra	2001	Geneva Observatory	0.73 AU	221.6s	Gas Giant	0.88 (Jupiter = 1)	0.54	Radial Velocity (or Doppler Spectroscopy)	89
HD 114783 b	HD 114783	Virgo	2001	Keck Precision Velocity Survey	1.2 AU	501s	Gas Giant	0.99 (Jupiter = 1)	0.1	Radial Velocity (or Doppler Spectroscopy)	72
HD 142 b	HD 142	Phoenix	2001	Anglo-Australian Planet Search	0.98 AU	338s	Gas Giant	1.36 (Jupiter = 1)	0.37	Radial Velocity (or Doppler Spectroscopy)	67
HD 4203 b	HD 4203	Pisces	2001	Keck Precision Velocity Survey	1.09 AU	406s	Gas Giant	1.64 (Jupiter = 1)	0.53	Radial Velocity (or Doppler Spectroscopy)	252
HD 68988 b	HD 68988	Ursa Major	2001	Keck Precision Velocity Survey	0.071 AU	6.276s	Hot Jupiter	1.9 (Jupiter = 1)	0.14	Radial Velocity (or Doppler Spectroscopy)	189
HD 213240 b	HD 213240	Grus	2001	Geneva Observatory	1.6 AU	759s	Gas Giant	3.7 (Jupiter = 1)	0.31	Radial Velocity (or Doppler Spectroscopy)	133

APPENDIX 6 - EXTRASOLAR PLANETS

Planet	Star	Constellation	Year of Discovery	Discovered By	Orbital Radius	Orbit Period (seconds)	Planet Type	Planet Mass	Orbit Eccentricity	Method of Detection	Star Distance from Sun (Light Years)
47 Ursae Majoris c	47 Ursae Majoris	Ursa Major	2001	University of California Planet Search	3.73 AU	2594s	Gas Giant	0.76 (Jupiter = 1)	0.1	Radial Velocity (or Doppler Spectroscopy)	43
HD 23079 b	HD 23079	Reticulum	2001	Anglo-Australian Planet Search	1.48 AU	627.3s	Gas Giant	2.54 (Jupiter = 1)	0.06	Radial Velocity (or Doppler Spectroscopy)	113
HD 80606 b	HD 80606	Ursa Major	2001	ESO La Silla Observatory	0.439 AU	111.78s	Gas Giant	3.41 (Jupiter = 1)	0.927	Radial Velocity (or Doppler Spectroscopy)	58.38
HD 28185 b	HD 28185	Eridanus	2001	CORALIE survey for southern extrasolar planets	1 AU	385s	Gas Giant	5.7 (Jupiter = 1)	0.06	Radial Velocity (or Doppler Spectroscopy)	128
HD 178911 b	HD 178911	Lyra	2001	Geneva Observatory	0.32 AU	71.487s	Gas Giant	6.292 (Jupiter = 1)	0.124	Radial Velocity (or Doppler Spectroscopy)	152
HD 106252 b	HD 106252	Virgo	2001	Geneva Observatory	2.61 AU	1500s	Gas Giant	6.81 (Jupiter = 1)	0.54	Radial Velocity (or Doppler Spectroscopy)	122
HD 33636 b	HD 33636	Orion	2001	Keck Precision Velocity Survey	2.62 AU	1553s	Gas Giant	7.71 (Jupiter = 1)	0.39	Radial Velocity (or Doppler Spectroscopy)	124
HD 39091 b	HD 39091	Mensa	2001	Anglo-Australian Planet Search	3.34 AU	2115.3s	Gas Giant	10.37 (Jupiter = 1)	0.62	Radial Velocity (or Doppler Spectroscopy)	67
HD 141937 b	HD 141937	Libra	2001	Geneva Observatory	1.49 AU	658.8s	Gas Giant	9.7 (Jupiter = 1)	0.4	Radial Velocity (or Doppler Spectroscopy)	109
Iota Draconis b	Iota Draconis	Draco	2002	Lick Observatory		547.5s	Gas Giant	8.7 (Jupiter = 1)	0.7	Radial Velocity (or Doppler Spectroscopy)	100
HD 41004A b	HD 41004A	Phoenix	2002	Geneva Observatory	1.3 AU	655s	Gas Giant	2.3 (Jupiter = 1)	0.39	Radial Velocity (or Doppler Spectroscopy)	138.5
HD 47536 b	HD 47536	Canis Major	2002	European Southern Observatory		712.13s	Gas Giant	> 4.96 (Jupiter = 1)	0.2	Radial Velocity (or Doppler Spectroscopy)	400

APPENDIX 6 - EXTRASOLAR PLANETS

Planet	Star	Constellation	Year of Discovery	Discovered By	Orbital Radius	Orbit Period (seconds)	Planet Type	Planet Mass	Orbit Eccentricity	Method of Detection	Star Distance from Sun (Light Years)
HD 136118 b	HD 136118		2002	Lick Observatory	2.335 AU	1209.6s	Gas Giant	11.9 (Jupiter = 1)	0.366	Radial Velocity (or Doppler Spectroscopy)	170
HD 160691 c	HD 160691	Ara	2002	Anglo-Australian Planet Search	2.3 AU	1300s	Gas Giant	1 (Jupiter = 1)	0.8	Radial Velocity (or Doppler Spectroscopy)	49
HD 49674 b	HD 49674		2002	Keck Observatory	0.0568 AU	4.948s	Hot Jupiter	0.12 (Jupiter = 1)	0	Radial Velocity (or Doppler Spectroscopy)	132.7
HD 108874 b	HD 108874		2002		1.07 AU	401s	Gas Giant	1.65 (Jupiter = 1)	0.2	Radial Velocity (or Doppler Spectroscopy)	223.31
HD 128311 b	HD 128311		2002	Keck Observatory	1.01 AU	414s	Gas Giant	2.63 (Jupiter = 1)	0.21	Radial Velocity (or Doppler Spectroscopy)	54
HD 72659 b	HD 72659		2002	Keck Observatory	3.24 AU	2185s	Gas Giant	2.55 (Jupiter = 1)	0.18	Radial Velocity (or Doppler Spectroscopy)	163
HD 40979 b	HD 40979		2002	Lick and Keck observatories	0.818 AU	260s	Gas Giant	3.16 (Jupiter = 1)	0.26		107.58
HD 114386 b	HD 114386		2002	Geneva Observatory	1.62 AU	872s	Gas Giant	0.99 (Jupiter = 1)	0.28	Radial Velocity (or Doppler Spectroscopy)	91.28
HD 150706 b	HD 150706		2002	Geneva Observatory	0.82 AU	264s	Gas Giant	1 (Jupiter = 1)	0.38	Radial Velocity (or Doppler Spectroscopy)	88.7
HD 147513 b	HD 147513		2002	Geneva Observatory	1.26 AU	540.4s	Gas Giant	1 (Jupiter = 1)	0.52	Radial Velocity (or Doppler Spectroscopy)	42
HD 20367 b	HD 20367		2002	Geneva Observatory	1.25 AU	500s	Gas Giant	1.07 (Jupiter = 1)	0.23	Radial Velocity (or Doppler Spectroscopy)	88
HD 30177 b	HD 30177		2002	Anglo-Australian Observatory	2.6 AU	1620s	Gas Giant	7.7 (Jupiter = 1)	0.22	Radial Velocity (or Doppler Spectroscopy)	179
HD 196050 b	HD 196050		2002	Anglo-Australian Observatory	2.5 AU	1289s	Gas Giant	3 (Jupiter = 1)	0.28	Radial Velocity (or Doppler Spectroscopy)	152

APPENDIX 6 - EXTRASOLAR PLANETS

Planet	Star	Constellation	Year of Discovery	Discovered By	Orbital Radius	Orbit Period (seconds)	Planet Type	Planet Mass	Orbit Eccentricity	Method of Detection	Star Distance from Sun (Light Years)
HD 23596 b	HD 23596		2002	Geneva Observatory	2.72 AU	1558s	Gas Giant	7.19 (Jupiter = 1)	0.314	Radial Velocity (or Doppler Spectroscopy)	169.5
Gliese 777A b	Gliese 777A		2002	Geneva Observatory	3.65 AU	2613s	Gas Giant	1.15 (Jupiter = 1)	-1	Radial Velocity (or Doppler Spectroscopy)	51.8
55 Cancri c	55 Cancri	Cancer	2002	Lick Observatory	0.24 AU	44.28s	Gas Giant	0.21 (Jupiter = 1)	0.34	Radial Velocity (or Doppler Spectroscopy)	44
55 Cancri d	55 Cancri	Cancer	2002	Lick Observatory	5.9 AU	5360s	Gas Giant	4.05 (Jupiter = 1)	0.16	Radial Velocity (or Doppler Spectroscopy)	44
HD 37124 c	HD 37124	Taurus	2002	Keck Observatory	2.95 AU	1942s	Gas Giant	1.01 (Jupiter = 1)	0.4	Radial Velocity (or Doppler Spectroscopy)	107
HD 12661 c	HD 12661	Aries	2002	Keck and Lick observatories	2.61 AU	1407s	Gas Giant	1.66 (Jupiter = 1)	0.224	Radial Velocity (or Doppler Spectroscopy)	121
HD 38529 c	HD 38529	Orion	2002	Keck and Lick observatories	3.51 AU	2189.5s	Gas Giant	11.3 (Jupiter = 1)	0.34	Radial Velocity (or Doppler Spectroscopy)	138
HD 114729 b	HD 114729		2002	Keck Observatory	2.08 AU	1136s	Gas Giant	0.9 (Jupiter = 1)	0.33		114
HD 216437 b	HD 216437		2002	Anglo-Australian Telescope	2.7 AU	1294s	Gas Giant	2.1 (Jupiter = 1)	0.34	Radial Velocity (or Doppler Spectroscopy)	135
HD 73526 b	HD 73526		2002	Anglo-Australian Planet Search	0.66 AU	190.5s	Gas Giant	3 (Jupiter = 1)	0.34	Radial Velocity (or Doppler Spectroscopy)	322.74
HD 76700 b	HD 76700		2002	Anglo-Australian Planet Search	0.049 AU	4s	Hot Jupiter	0.197 (Jupiter = 1)	0	Radial Velocity (or Doppler Spectroscopy)	194.6
HD 2039 b	HD 2039		2002	Anglo-Australian Planet Search	2.2 AU	1190s	Gas Giant	5.1 (Jupiter = 1)	0.69	Radial Velocity (or Doppler Spectroscopy)	292.7
Tau I Gruis b	Tau I Gruis	Grus (the crane)	2002	Anglo-Australian Planet Search	2.5 AU		Gas Giant	1.2 (Jupiter = 1)	-1	Radial Velocity (or Doppler Spectroscopy)	100

APPENDIX 6 - EXTRASOLAR PLANETS

Planet	Star	Constellation	Year of Discovery	Discovered By	Orbital Radius	Orbit Period (seconds)	Planet Type	Planet Mass	Orbit Eccentricity	Method of Detection	Star Distance from Sun (Light Years)
gamma Cephei b	gamma Cephei	Cepheus	2002	McDonald Observatory	2.1 AU	903s	Gas Giant	1.76 (Jupiter = 1)	0.2	Radial Velocity (or Doppler Spectroscopy)	38.5
Epsilon Eridani c	Epsilon Eridani	Eridanus	2002	Alice Quillen, University of Rochester		102200s	Gas Giant	0.1 (Jupiter = 1)	-1		10.4
HD 216770 b	HD 216770	Pegasus	2003	Geneva Observatory	0.46 AU	118.3s	Gas Giant	0.7 (Jupiter = 1)	0.32	Radial Velocity (or Doppler Spectroscopy)	124
HD 104985 b	HD 104985	Draco	2003	Okayama Astrophysical Observatory	0.78 AU	198.2s	Gas Giant	6.3 (Jupiter = 1)	0.03	Radial Velocity (or Doppler Spectroscopy)	332.5
HD 70642 b	HD 70642	Puppis	2003	Anglo-Australian Telescope	3.3 AU	2231s	Gas Giant	2 (Jupiter = 1)	0.1	Radial Velocity (or Doppler Spectroscopy)	94.5
HD 74156 b	HD 74156	Hydra	2003	Geneva Observatory	0.276 AU	51.61s	Gas Giant	1.56 (Jupiter = 1)	0.649	Radial Velocity (or Doppler Spectroscopy)	210
HD 142415 b	HD 142415		2003	CORALIE Survey for extrasolar planets	1.05 AU	386.3s	Gas Giant	1.62 (Jupiter = 1)	0.5	Radial Velocity (or Doppler Spectroscopy)	111.5
HD 330075 b	HD 330075		2003	HARPS	0.044 AU	3.37s	Hot Jupiter	0.8 (Jupiter = 1)	-1	Radial Velocity (or Doppler Spectroscopy)	163.6
HD 3651 b	HD 3651	Pegasus	2003	California & Carnegie Planet Search Team	0.284 AU	62.23s	Gas Giant	0 (Jupiter = 1)	0.63	Radial Velocity (or Doppler Spectroscopy)	35.7
OGLE-TR 56	OGLE-TR-56	Sagittarius	2003	Harvard-Smithsonian Center for Astrophysics	0.0225 AU	1.2119s	Hot Jupiter	0.9 (Jupiter = 1)	-1	Transit Method	5,000
HD 73256 b	HD 73256	Canis Major	2003	CORALIE survey for extrasolar planets	0.037 AU	2.5486s	Hot Jupiter	1.85 (Jupiter = 1)	0.038	Radial Velocity (or Doppler Spectroscopy)	119
HD 10647 b	HD 10647	Phoenix	2003	Geneva Observatory	2.1 AU	1056s	Gas Giant	1.17 (Jupiter = 1)	0.32	Radial Velocity (or Doppler Spectroscopy)	56.4
HD 111232 b	HD 111232	Musca	2003	Geneva Observatory	2.07 AU	1138s	Gas Giant	7.8 (Jupiter = 1)	0.25	Radial Velocity (or Doppler Spectroscopy)	94.5

APPENDIX 6 - EXTRASOLAR PLANETS

Planet	Star	Constellation	Year of Discovery	Discovered By	Orbital Radius	Orbit Period (seconds)	Planet Type	Planet Mass	Orbit Eccentricity	Method of Detection	Star Distance from Sun (Light Years)
HD 169830 c	HD 169830	Sagittarius	2003	CORALIE Survey for Extrasolar Planets	3.6 AU	2102s	Gas Giant	4.04 (Jupiter = 1)	0.33	Radial Velocity (or Doppler Spectroscopy)	118
HD 65216 b	HD 65216		2003	CORALIE Survey for Extrasolar Planets	1.37 AU	613.1s	Gas Giant	1.21 (Jupiter = 1)	0.41	Radial Velocity (or Doppler Spectroscopy)	111.8
HD 8574 b	HD 8574	Pisces	2003	Geneva Observatory	0.76 AU	228.8s	Gas Giant	2.23 (Jupiter = 1)	0.4	Radial Velocity (or Doppler Spectroscopy)	144
OGLE-TR-111 b	OGLE-TR-111		2004	Optical Gravitational Lensing Experiment	0.047 AU	4s	Hot Jupiter	0.53 (Jupiter = 1)	0	Transit Method	4,875
HD 219449 b	HD 219449	Aquarius	2004	Lick Observatory	0.3 AU	182s	Gas Giant	2.9 (Jupiter = 1)	-1	Radial Velocity (or Doppler Spectroscopy)	146
HD 59686 b	HD 59686		2004	Lick Observatory	0.8 AU	303s	Gas Giant	6.5 (Jupiter = 1)	-1	Radial Velocity (or Doppler Spectroscopy)	299
OGLE-TR-132 b	OGLE-TR-132		2004	OGLE (Optical Gravitational Lensing Experiment)	1.15 AU	1.7s	Hot Jupiter	1.01 (Jupiter = 1)	-1	Transit Method	4,890
OGLE-TR-113 b	OGLE-TR-113		2004	Las Campanas Observatory, Chile	0.0228 AU	1.4s	Hot Jupiter	1.35 (Jupiter = 1)	0	Transit Method	4,890
OGLE 2003-BLG-235/ MOA 2003-BLG-53	OGLE 2003-BLG-235/MOA 2003-BLG-53		2004	Microlensing Observations in Astrophysics (MOA) and Optical Gravitational Lensing Experiment (OGLE)			Gas Giant	> 1.5 (Jupiter = 1)	-1	Gravitational Microlensing	17,000
HD 37605 b	HD 37605		2004	McDonald Observatory astronomers, using the Hobby-Eberly Telescope	0.26 AU	54.23s	Gas Giant	2.84 (Jupiter = 1)	0.737	Radial Velocity (or Doppler Spectroscopy)	139.8
TrES-1	GSC 02652-01324		2004		0.04 AU	3.03s	Hot Jupiter	0.75 (Jupiter = 1)	0	Transit Method	500
HD 160691 d	HD 160691	Ara	2004	European Southern Observatory HARPS instrument	0.09 AU	9.5s	Gas Giant	14 (Earth = 1)	0	Radial Velocity (or Doppler Spectroscopy)	49
55 Cancri e	55 Cancri	Cancer	2004	Hobby-Eberly Telescope	0.04 AU	3s	Unknown	18 (Earth = 1)	-1	Radial Velocity (or Doppler Spectroscopy)	44
GJ 436 b	GJ 436	Leo	2004	Keck Observatory	0.02 AU	2.6s	Unknown	0.067 (Jupiter = 1)	0.12	Radial Velocity (or Doppler Spectroscopy)	30

APPENDIX 6 - EXTRASOLAR PLANETS

Planet	Star	Constellation	Year of Discovery	Discovered By	Orbital Radius	Orbit Period (seconds)	Planet Type	Planet Mass	Orbit Eccentricity	Method of Detection	Star Distance from Sun (Light Years)
HD 154857 b	HD 154857	Apus	2004	Anglo-Australian Planet Search	1.11 AU	398.5s	Gas Giant	1.8 (Jupiter = 1)	0.51	Radial Velocity (or Doppler Spectroscopy)	222
HD 11768 b	HD 117618	Centaurus	2004	Anglo-Australian Planet Search	0.15 AU	25.8s	Hot Jupiter	0.16 (Jupiter = 1)	0.31	Radial Velocity (or Doppler Spectroscopy)	123
HD 102117	HD 102117	Carina	2004	Anglo-Australian Planet Search	0.13 AU	20.8s	Hot Jupiter	0.18 (Jupiter = 1)	0.08	Radial Velocity (or Doppler Spectroscopy)	99
HD 208487 b	HD 208487	Grus	2004	Ango-Australian Planet Search	0.5 AU	129s	Hot Jupiter	0.41 (Jupiter = 1)	0.47	Radial Velocity (or Doppler Spectroscopy)	145.8
HD 88133 b	HD 88133	Leo	2004	N2K Consortium	0.046 AU	3.415s	Gas Giant	0.29 (Jupiter = 1)	0.11	Radial Velocity (or Doppler Spectroscopy)	241.4
HD 202206 c	HD 202206	Capricornus	2004	CORALIE Survey for Southern Extrasolar Planets	2.55 AU	1383.4s	Gas Giant	2.44 (Jupiter = 1)	-1	Radial Velocity (or Doppler Spectroscopy)	150
HD 183263 b	HD 183263	Aquila	2005	Keck Observatory	1.52 AU	634s	Gas Giant	3.69 (Jupiter = 1)	0.38	Radial Velocity (or Doppler Spectroscopy)	172
HD 117207 b	HD 117207	Centaurus	2005	Keck Observatory	3.78 AU	2627.08s	Gas Giant	2.06 (Jupiter = 1)	0.16	Radial Velocity (or Doppler Spectroscopy)	107
HD 188015 b	HD 188015	Vulpecula	2005		1.19 AU	456.46s	Gas Giant	1.26 (Jupiter = 1)	0.15	Radial Velocity (or Doppler Spectroscopy)	171
HD 45350 b	HD 45350	Auriga	2005		1.77 AU	890.76s	Gas Giant	0.98 (Jupiter = 1)	0.78	Radial Velocity (or Doppler Spectroscopy)	159
HD 99492 b	HD 99492	Leo	2005		0.119 AU	17.038s	Gas Giant	0.122 (Jupiter = 1)	0.05	Radial Velocity (or Doppler Spectroscopy)	58.5
HD 2638 b	HD 2638	Andromeda	2005		0.044 AU	3.4s	Hot Jupiter	0.48 (Jupiter = 1)	-1	Radial Velocity (or Doppler Spectroscopy)	175

APPENDIX 6 - EXTRASOLAR PLANETS

Planet	Star	Constellation	Year of Discovery	Discovered By	Orbital Radius	Orbit Period (seconds)	Planet Type	Planet Mass	Orbit Eccentricity	Method of Detection	Star Distance from Sun (Light Years)
HD 27894 b	HD 27894	Hydrus	2005	High Accuracy Radial velocity Planetary Search (HARPS)	0.122 AU	18s	Hot Jupiter	0.62 (Jupiter = 1)	0.049	Radial Velocity (or Doppler Spectroscopy)	137.7
HD 63454 b	HD 63454	Mensa	2005	High Accuracy Radial velocity Planetary Search (HARPS)	0.036 AU	2.8s	Hot Jupiter	0.38 (Jupiter = 1)	-1	Radial Velocity (or Doppler Spectroscopy)	116.3
HD 93083 b	HD 93083	Antlia	2005	High Accuracy Radial velocity Planetary Search (HARPS)	0.477 AU	143.6s	Gas Giant	0.37 (Jupiter = 1)	0.14	Radial Velocity (or Doppler Spectroscopy)	97
HD 102117 b	HD 102117	Centaurus	2005	High Accuracy Radial velocity Planetary Search (HARPS)	0.302 AU	70.46s	Gas Giant	0.3 (Jupiter = 1)	0.11	Radial Velocity (or Doppler Spectroscopy)	99
HD 142022A b	HD 142022A	Octans	2005		2.8 AU	1923s	Gas Giant	4.4 (Jupiter = 1)	0.57	Radial Velocity (or Doppler Spectroscopy)	116.6
2M1207 b	2M1207 (brown dwarf)		2005		55 AU		Gas Giant	5 (Jupiter = 1)	-1	Direct Imaging	227.5
OGLE-2005-BLG-071	OGLE-2005-BLG-071		2005				Gas Giant	> 3 (Jupiter = 1)	-1	Gravitational Microlensing	15,000
Gliese 876 d	Gliese 876	Aquarius	2005		0.02 AU	1.9s	Unknown	7.5 (Earth = 1)	-1	Radial Velocity (or Doppler Spectroscopy)	15
HD 108874 c	HD 108874		2005		2.68 AU	1605.8s	Gas Giant	0 (Jupiter = 1)	0.25	Radial Velocity (or Doppler Spectroscopy)	223.31
HD 128311 c	HD 128311		2005		1.02 AU	919s	Gas Giant	3.21 (Jupiter = 1)	0.17	Radial Velocity (or Doppler Spectroscopy)	54
Gliese 777A c	Gliese 777A		2005		0.128 AU	17s	Gas Giant	18.1 (Earth = 1)	0.01	Radial Velocity (or Doppler Spectroscopy)	51.8
HD 217107 c	HD 217107	Pisces	2005	California Planet Search Team	4.3 AU	3150s	Gas Giant	2.1 (Jupiter = 1)	0.55	Radial Velocity (or Doppler Spectroscopy)	121
HD 37124 d	HD 37124	Taurus	2005		3.19 AU	2295s	Gas Giant	0.66 (Jupiter = 1)	0.2	Radial Velocity (or Doppler Spectroscopy)	107

APPENDIX 6 - EXTRASOLAR PLANETS

Planet	Star	Constellation	Year of Discovery	Discovered By	Orbital Radius	Orbit Period (seconds)	Planet Type	Planet Mass	Orbit Eccentricity	Method of Detection	Star Distance from Sun (Light Years)
HD 149026 b	HD 149026	Hercules	2005		0.042 AU	2.9s	Gas Giant	0.36 (Jupiter = 1)	-1	Radial Velocity (or Doppler Spectroscopy)	256
HD 188753 Ab	HD 188753		2005			3.3s	Hot Jupiter	1.14 (Jupiter = 1)	0	Radial Velocity (or Doppler Spectroscopy)	149
HD 189733 b	HD 189733	Vulpecula	2005		0.03 AU	2.2s	Hot Jupiter	1.15 (Jupiter = 1)	0	Radial Velocity (or Doppler Spectroscopy)	63
HD 81040 b	HD 81040	Leo	2005		1.94 AU	1001.7s	Gas Giant	6.86 (Jupiter = 1)	0.526	Radial Velocity (or Doppler Spectroscopy)	106
Gliese 581 b	Gliese 581	Libra	2005		0.041 AU	5.4s	Gas Giant	17 (Earth = 1)	0	Radial Velocity (or Doppler Spectroscopy)	20.5
HD 102195 b	HD 102195	Leo	2006		0.049 AU	4.1s	Gas Giant	0.48 (Jupiter = 1)	0.06	Radial Velocity (or Doppler Spectroscopy)	94
OGLE-05-390L b	OGLE-05-390L	Sagittarius	2006		2.6 AU	3800s	Unknown	5.5 (Earth = 1)	-1	Gravitational Microlensing	21,000
HD 73526 c	HD 73526		2006	Anglo-Australian Planet Search	1.05 AU	377.8s	Gas Giant	1.5 (Jupiter = 1)	-1	Radial Velocity (or Doppler Spectroscopy)	322.74
HD 187085 b	HD 187085	Sagittarius	2006		2.05 AU	986s	Gas Giant	0.75 (Jupiter = 1)	-1	Radial Velocity (or Doppler Spectroscopy)	146.7
HD 20782 b	HD 20782	Fornax	2006		1.36 AU	585.86s	Gas Giant	1.8 (Jupiter = 1)	-1	Radial Velocity (or Doppler Spectroscopy)	117
OGLE-05-169L b	OGLE-05-169L		2006				Terrestrial	13 (Earth = 1)	-1	Gravitational Microlensing	9,000
Gliese 876 c	Gliese 876	Aquarius	Unknown		0.13 AU	30.1 days	Gas Giant	0.56 (Jupiter = 1)	0.27	Radial Velocity (or Doppler Spectroscopy)	15
HD 50554 b	HD 50554	Gemini	Unknown	Geneva Observatory	2.38 AU	1279 days	Gas Giant	4.9 (Jupiter = 1)	0.42	Radial Velocity (or Doppler Spectroscopy)	101
HD 74156 c	HD 74156	Hydra	Unknown	Geneva Observatory	4.47 AU	2300 days	Gas Giant	$>$ 7.5 (Jupiter = 1)	0.359	Radial Velocity (or Doppler Spectroscopy)	210

Space Exploration 2007
Harvey
HB 0-387-33330

In Search of Dark Matter
Freeman/McNamara
PB 0-387-27616

Searching for Water in the Universe
Encrenaz
PB 0-387-34174

The Sky at Einstein's Feet
Keel
PB 0-387-26130

www.springer.com/space www.praxis-publishing.co.uk

LIFE IN THE
SOLAR SYSTEM
AND
BEYOND
BARRIE W. JONES
Springer
The BIG
BANG
A View from the 21st Century
Springer
PRAXIS
David M. Harland

IF THE UNIVERSE IS TEEMING WITH ALIENS
WHERE IS EVERYBODY?
FIFTY SOLUTIONS TO THE FERMI PARADOX
AND THE PROBLEM OF EXTRATERRESTRIAL LIFE

THE END